The Way to
Vibrant Health

The Way to Vibrant Health

A MANUAL OF BIOENERGETIC EXERCISES

Alexander Lowen, M.D.
Leslie Lowen

*Illustrations by Walter Skalecki who has been
associated with bioenergetics for years*

HARPER COLOPHON BOOKS
Harper & Row, Publishers
New York, Hagerstown, San Francisco, London

Designed by Eve Kirch Callahan

First HARPER COLOPHON edition published 1977

LIBRARY OF CONGRESS CATALOG CARD NUMBER: 76–15320

ISBN: 0– 06– 090542– 5

78 79 80 10 9 8 7 6 5 4 3

Contents

Part I: The Basics of Bioenergetics

Part II: The Exercises

Part III: Setting Up a Regular Schedule

PART I

The Basics of Bioenergetics

Introduction: What Is Bioenergetics?

Bioenergetics is a way of understanding personality in terms of the body and its energetic processes. These processes, namely, the production of energy through respiration and metabolism and the discharge of energy in movement, are the basic functions of life. How much energy one has and how one uses his energy determine how one responds to life situations. Obviously, one can cope more effectively if one has more energy, which can be freely translated into movement and expression.

Bioenergetics is also a form of therapy that combines work with the body and the mind to help people resolve their emotional problems and realize more of their potential for pleasure and joy in living. A fundamental thesis of bioenergetics is that body and mind are functionally identical: that is, what goes on in the mind reflects what is happening in the body and vice versa. The relationship between these three elements, body, mind, and energetic processes, is best expressed by a dialectical formulation as shown in the following diagram.

As we all know, mind and body can influence each other. What one thinks can affect how one feels. The converse is equally true. This interaction, however,

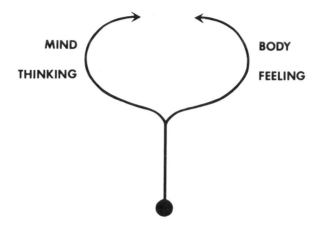

Fig. 1. Energetic processes

is limited to the conscious or superficial aspects of the personality. On a deeper level, that is, on the unconscious level, both thinking and feeling are conditioned by energy factors. For example, it is almost impossible for a depressed person to lift himself out of his depression by thinking positive thoughts. This is because his energy level is depressed. When his energy level is raised through deep breathing (his breathing was depressed along with all other vital functions) and the release of feeling, the person comes out of the depressed state.*

The energetic processes of the body are related to the state of aliveness of the body. The more alive one is, the more energy one has and vice versa. Rigidity or chronic tension diminishes one's aliveness and decreases one's energy. At birth, an organism is in its most alive, most fluid state; at death, rigidity is total, rigor mortis. We cannot avoid the rigidity that comes with age. What we can avoid is the rigidity due to chronic muscular tensions resulting from unresolved emotional conflicts.

*See my book *Depression and the Body* (Baltimore: Penguin Books, 1973).

Every stress produces a state of tension in the body. Normally the tension disappears when the stress is relieved. Chronic tensions, however, persist after the provoking stress has been removed as an unconscious bodily attitude or muscular set. Such chronic muscular tensions disturb emotional health by decreasing an individual's energy, restricting his motility (the natural spontaneous play and movement of the musculature), and limiting his self-expression. It becomes necessary then to relieve this chronic tension if the person is to regain his full aliveness and emotional well-being.

The body work of bioenergetics includes both manipulative procedures and special exercises. The manipulative procedures consist of massage, controlled pressure, and gentle touching to relax contracted muscles. The exercises are designed to help a person get in touch with his tensions and release them through appropriate movement. It is important to know that every contracted muscle is blocking some movement. These exercises have been developed in the course of more than twenty years of therapeutic work with patients. They are done in therapy sessions, in classes, and at home. People who do these exercises report a positive effect upon their energy, their mood, and their work. The authors do them regularly to promote their own well-being. Wherever we have introduced these exercises, for example, at workshops for professional people, the response has been enthusiastic. We are constantly asked for a list and description of the exercises. This manual is our response to that demand.

We wish to emphasize at the outset that these exercises are not a substitute for therapy. They will not resolve deep emotional problems, which generally require competent professional help. Very often people who are not in therapy and who do these exercises decide that they need and want such help to work through problems that may have risen to awareness during the course of these exercises. But whether or not you are in therapy, the regular performance of these exercises will help you significantly to increase your aliveness and capacity for pleasure.

These exercises can help you gain more self-possession, with all that this term implies. They will do this by (1) increasing the vibratory state of your body, (2) grounding you in your legs and body, (3) deepening your respiration, (4) sharpening your self-awareness, and (5) enlarging your self-expression. They can also improve your figure, heighten your sexual feelings, and promote your self-confidence. However, they are exercises, not skills, and much depends on what you put into them. If you do them mechanically, you will get little out of them. If you do them compulsively, their value will diminish. If you do them competitively, you will prove nothing. However, if you do them with care for and interest in your body, the benefits will astonish you.

1

Vibration and Motility

As we have indicated, bioenergetics is the vibrant way to health and the way to vibrant health. By "vibrant health" we mean not merely the absence of illness but the condition of being fully alive. Vibrantly alive is perhaps a more accurate term, for vibration is the key to aliveness. By increasing the vibratory state of the body through these exercises, a person is helped to approach this quality of health.

A healthy body is in a constant state of vibration whether awake or asleep. Look at a sleeping infant and you will see fine tremors pass over the surface of his body. You may observe small twitches in different parts, the face especially, but also the arms and legs. We adults sometimes experience these tremors or twitches, too. A living body is in constant motion; only in death is it truly still. This inherent motility of a living body, which is the basis of its spontaneous activity, results from a state of inner excitement that is continually erupting on the surface in movement. When the excitement mounts, there is more movement; when it falls, the body becomes quieter.

As the vibratory state of the body increases in a coordinated manner, pulsatory waves develop and spread through the body. We are familiar with these waves in the beat of the heart that pulses through the arteries and in the peristaltic movement of the intestines, which is a pulsatory wave. But we do not often experience the pulsatory waves that flow through the whole body in states of full relaxation or intense feeling. In full relaxation, respiratory waves pass through the body with each inspiration and expiration (inhaling and exhaling). In states of strong emotion, waves of feeling sweep through the body. Similar pulsatory waves occur in the climax of the sexual act. Usually, however, we do not allow ourselves to relax fully, breathe deeply, or feel intensely.

Vibration is due to an energetic charge in the musculature and is analogous to the vibration occurring in an electrical wire when a current passes through it. The lack of vibration is an indication that the current of excitation or charge is absent or greatly reduced. One can get a clearer picture of this phenomenon by considering what happens to a car when the ignition is turned on. As it starts up it goes into a strong vibration, which then settles down to a steady hum. This hum (or vibration) will continue as long as the engine is running. Should the engine stop while the car is moving, one immediately senses that it has gone dead by the absence of the hum.

The quality of the vibration in a car or a person's body tells us what shape it is in. When the car shakes or the vibrations are rough, we sense that something is amiss. In a body, gross vibrations are a sign that the excitation or charge is not flowing freely. Just as rapids in a river denote that rocks or other obstacles impede what would otherwise be the smoothness of its course, so too gross vibrations denote that the current of excitation is flowing through muscles that are spastic or in a state of chronic tension. When the tensions are released or the muscle relaxes, the vibrations become finer, hardly perceptible on the surface yet experienced as a delightful purr. Still, it is better to shake than not to vibrate at all. Then, too, there are conditions when a body will shake because of an extremely intense charge. For instance, we shake with anger or tremble with fear, or convulse with sobs and pulsate with love; but regardless of the emotion, we are fully alive in these states.

In the course of bioenergetic work, a person's body is brought into a state of vibration through the special exercises described in this manual. The objective is to keep the vibrations going at a fine and steady purr as the excitement builds or the stress increases. In effect, one increases the body's tolerance for excitation and for pleasure. To accomplish this the ego has to be securely anchored in the body, identified with it, and unafraid to go with the body's involuntary responses. The end result is a person whose movements and behavior have a high degree of spontaneity and yet are coordinated and effective: the quality of natural grace.

During this process there is a corresponding change in a person's thinking and attitudes. When the vibrations pass fully through the body, a person feels connected and integrated, all of a piece. Many patients have commented on this reaction. The feeling of unity and integrity leads to a natural sincerity in thought and action. If a person develops bodily grace, he develops the corresponding psychological attitude of being gracious. Such persons are not only vibrantly alive, they are radiantly alive.

Bioenergetic analysis is the name for bioenergetic therapy. In bioenergetic therapy a person is helped to get in touch with himself through his body. By using the exercises described in this manual, the person begins to sense how he inhibits or blocks the flow of excitation in his body; how he has limited his breathing, restricted his movements and reduced his self-expression; in other words, how he has decreased his aliveness. The analytic part of the therapy helps him understand the *why* of these mostly unconscious inhibitions and blocks in terms of his childhood experiences. He is helped and encouraged to accept and express the suppressed feelings in the controlled setting of the therapeutic situation.

The goal of the therapy is an alive body, one capable of fully experiencing the pleasures and pains, the joys and sorrows of life. The more alive we are, the more we can tolerate a heightened excitement in our daily lives and in sex. Analysis of repressed conflicts, release of suppressed feelings, and dissolution of chronic muscular tensions and blocks have the purpose of increasing a person's capacity for pleasure.

The pleasure of being fully alive is anchored in the vibratory state of the body. It is perceived in the full pulsatory expansion and contraction of the organism and its component organ systems, the respiratory, circulatory, and digestive systems, for example. It is felt as streaming sensations in the body reflecting the flow of excitation. It is the sweet melting sensation of sexual desire, the flash of intuition, the longing for closeness and contact, and the throb of excitement.

Vibratory activity is, as we noted earlier, a manifestation of the inherent motility of the organism, which is also responsible for spontaneous actions, emotional releases, and internal functioning. This inherent motility is not under the control of the ego or will; it is involuntary. An alive body pulses and vibrates. Naturally, as we become older our bodies become more and more static until they reach the absolute stillness of death. But the premature loss of motility is pathological. This happens, for instance, when we become depressed. Depression is a pathological decrease in the vital functioning of the body, a diminution of motility, feeling, and responsiveness.

In addition to these involuntary movements, we also make many voluntary movements, consciously or semiconsciously, such as walking, talking, eating, and so on. In a healthy adult the two kinds of movement, the involuntary and the voluntary, are finely coordinated to produce behavior that is both graceful and effective. This is the way we would all like to be. But true grace cannot be learned. What one learns in a modeling school is how to be a mannequin, not a graceful, alive person. The pose may look attractive in a picture but it strikes one as stiff and awkward in real life, for it is achieved at the expense of the spontaneous motility of the body. One can only achieve grace by increasing the motility of the body, then fusing it with self-awareness to yield a high degree of self-possession. The mark of the graceful and gracious person is his self-possession.

One of the most fundamental exercises in bioenergetics is also the easiest and simplest. We use it to start the vibrations in the legs and to help the person sense them. It is also our basic grounding exercise. Doing it without any preliminary warm-up may or may not result in any vibrations. Young people generally respond quickly. Older persons, whose bodies are less charged and more rigid, might not experience them. However, their legs, too, can vibrate after they have done some of the other exercises that reduce their rigidity, deepen their breathing, and increase their energetic charge (amount of energy, excitation, or current in the body).

Exercise 1 / Basic vibratory and grounding exercise

Stand with feet about 10″ apart, toes slightly turned in so as to stretch some of the muscles of the buttocks. Bend forward and touch the floor with the fingers of both hands as in figure 2. The knees should be slightly bent. No weight should be on the hands; all the body weight is in the feet. Let the head drop as much as possible.

Breathe through your mouth easily and deeply. Make sure to keep breathing. (Forget about breathing through your nose for the time being.)

Let the weight of your body go forward so that it is on the balls of the feet. The heels can be slightly raised.

Straighten the knees slowly until the hamstring muscles at the back of the legs are stretched. However,

Fig. 2. Vibrating bent forward

the knees should not be fully straightened or locked.

Hold the position for about one minute.

• Are you breathing easily or are you holding your breath? No vibrations will occur if you stop breathing.

• Do you sense any vibratory activity in your legs? If not, try slowly bending the knee a little, then straighten it to the original position. Do this a number of times to get the muscles to relax.

• Are the vibrations fine or gross, smooth or jerky? In some cases people literally jump off the ground if they can't hold the excitation. Did this happen to you?

We will ask you to try this exercise again after reading the next chapter.

2

Grounding

You may have noticed if you did the exercise from the preceding chapter that the vibrations in your legs occur when you feel your feet pressing on the ground. The feeling contact between the feet and the ground is known in bioenergetics as *grounding*. This denotes a flow of excitation through the legs into the feet and ground. One is then connected to the ground, not "up in the air" or "hung-up." There are, of course, different degrees of feeling contact with the ground depending upon how fully the feet "touch" the ground. People vary widely in this.

To be grounded is another way of saying that a person has his feet on the ground. It can also be extended to mean that a person knows where he stands and therefore that he knows who he is. Being grounded, a person has "standing," that is, he is "somebody." In a broader sense grounding represents an individual's contact with the basic realities of his existence. He (or she) is rooted in the earth, identified with his body, aware of his sexuality, and oriented toward pleasure. These qualities are lacking in the person who is "up in the air" or in his head instead of in his feet.

Grounding involves getting a person to "let down," to lower his center of gravity, to feel closer to the earth. The immediate result is to increase his sense of security. He feels the ground under him and his feet resting on it. When a person becomes highly charged or excited, he tends to go upward, to fly, or to fly off. In this condition, despite a sense of excitation, or elation, there is always an element of anxiety and danger, namely, the danger of falling. This is equally true where one is off the ground as in an airplane. It is resolved when the person is back safe on the ground, physically or emotionally.

The direction downward is the way to the pleasure of release or discharge. It is the way to sexual satisfaction. Persons who are afraid to let down are blocked in their ability to surrender fully to the sexual discharge and fail to experience full orgastic satisfaction. Letting go means letting down, for we are unconsciously holding ourselves up all the time. We are afraid to fall, afraid to fail, and therefore afraid to let go and give in to our feelings.

Mabel Elsworth Todd in her book *The Thinking Body,* first published in 1937, made this observation: "Man has become absorbed with the upper portions of the body in intellectual pursuits and in the development of skills of hand and speech. This, in addition to false notions regarding appearances or health, has transferred his sense of power from the base to the top of his structure. In thus using the upper part of the body for power reactions he has reversed the animal usage and has to a great extent lost both the fine sensory capacity of the animal and its control of power centered in the lower spinal and pelvic muscles."*

In a broad sense, grounding aims at helping a person become more fully identified with his animal nature, which, of course, includes his sexuality. The lower half of the body is much more animallike in its functions (locomotion, defecation, and sexuality) than the upper half (thinking, speaking, and manipulating the environment). These functions are more instinctive and less subject to conscious control. But it is in our animal nature that the qualities of rhythm and

*Mabel Elsworth Todd, *The Thinking Body* (New York: Paul B. Hober, Inc., 1937), p. 160. Republished by Dance Horizons, Inc., New York.

grace reside. Any movement that flows freely from the lower part of the body has these qualities. When we pull ourselves up and away from the lower half of the body, we lose much of our natural rhythmicity and grace.

This upward displacement can be reversed through the bioenergetic grounding exercises. As the body's center of gravity drops into the pelvis with the feet serving as energetic supports, one can sense the self centered in the lower abdomen.

The importance of being centered in the lower abdomen or belly is recognized by most Orientals. The Japanese, for example, have a word, *hara,* which means the belly and also the quality of being a person who is centered in this region. The exact point, according to Durckheim, is 2″ below the navel. If a person is centered at this point he is said to have *hara,* that is, he is balanced both psychologically and physically. The balanced person is calm and at ease; all his movements are effortless yet masterful. Durckheim writes: "When a man possesses fully developed Hara he has the strength and precision to perform actions which otherwise he could never achieve even with the most perfect technique, the closest attention or the strongest will-power. 'Only what is done with Hara succeeds completely.' "* The disciplines of Zen archery, flower arranging, and the tea ceremony are designed to help a person attain *hara.*

Most Westerners are centered in the upper part of the body, mainly in the head. We recognize the head as the focus of the ego, the center of consciousness and deliberate behavior. In contrast, the lower or pelvic center where *hara* resides is the center for the unconscious or instinctive life. Let us say that it is man's animal center, as Todd suggests. When we realize that no more than 10 percent of our movements are consciously directed and that 90 percent are unconscious, the importance of this center becomes evident.

An analogy will make this clear. Think of a horse and rider. The rider with his

*Karlfried Durckheim, *Hara, The Vital Center of Man* (London: George Allen & Unwin, Ltd., 1962), p. 46.

conscious control of direction and speed functions like the ego; the horse provides the lower center, the power, and surefootedness to carry the rider where he wants to go. Should the rider become unconscious, the horse would in most cases bring him safely home. But should the horse break down, the rider would be virtually helpless. The best he might do is walk to his destination.

The belly is literally the seat of life. The body sits in the pelvic basket. Through the pelvis, one has contact with the sexual organs and the legs. It is also in the belly that the individual is conceived, and from the belly he emerges downward into the light of day. The loss of contact with this vital center imbalances a person and leads to anxiety and insecurity.

There are two commandments that, if observed, help you become and stay grounded. The first is to keep your knees slightly flexed at all times. Locking the knees when standing turns the whole lower part of the body from the hips down into a rigid structure, which then functions as a mechanical support or a mechanical means of locomotion. It prevents one from flowing into and identifying with the lower part of the body.

The knees are the shock absorbers of the body. When pressure is exerted on a person, the knees flex, allowing the force to be transmitted through the body and into the ground. If the knees are locked, the force is trapped in the lower back, producing a stress condition that will result in lower back trouble. We are always advised to keep our knees bent when lifting heavy objects. We fail to realize that psychological pressures are the equivalent of physical weights to the body. If we attempt to support these pressures with locked knees, we take their force in our lower back.*

*A fuller discussion of the dynamics of stress will be found in my book *Bioenergetics* (New York: Coward, McCann & Geoghegan, 1975).

Exercise 2 / Flexing the knees

Stand with your feet about 8″ apart in your normal position. Observe whether your knees are locked or bent, whether your feet are parallel or turned outward, whether your weight is forward on the balls of your feet or backward on your heels.

Now bend your knees slightly. Turn your feet so that they are absolutely parallel. Pitch your weight forward without raising your heels so that it rests on the balls of your feet. Slowly bend and straighten the knees six times, and then hold the position for about thirty seconds, breathing easily.

- Does this position feel unnatural to you? If it does, you have not been standing correctly.
- Do your legs feel shaky? Do you feel insecure on them?
- Do you have a better sense of your feet on the ground?
- Are you aware of the flexibility the knees provide when they are not locked or held tight?

The second commandment is to let the belly out. Many people find this hard to do at first. It violates their image of correct posture and good appearance. They have been brainwashed with the dictum for proper bearing: "belly in, chest out, shoulders up." Perhaps this bearing is proper for a soldier who must function like an automaton, but it is the epitome of rigidity. It denies a person autonomy, spontaneity, and sexuality. The sucked-in belly makes abdominal breathing very difficult and forces one to overinflate the chest to get enough air. The continued overinflation of the chest is one of the factors responsible for emphysema. In the next chapter we will describe the healthy or correct breathing pattern more fully. As we shall see, it is dependent upon a relaxed abdominal musculature.

By holding your belly in and your shoulders up, you are using a lot of energy to fight your basic animal nature. And you will not succeed, although you will tire yourself. If someone ordered you to walk around holding your right hand up like the Statue of Liberty as a symbol of freedom, you would regard such a pose as an unnecessary strain. This is equally true of any posture that is forced or willed.

It is work to assume any body attitude that requires effort, unnecessary and wasteful work which only serves to create an image.

Letting the belly out seems to offend women especially. They see it as sloppy and unattractive. Their image of feminine beauty is the *Playboy* bunny, with her tightly sucked-in belly and stuck-out breasts. This is supposed to be sexually exciting to men. Perhaps it is to some men, those who are repelled and afraid of a woman with a belly whom they see as a mother figure. However, a belly is an indication of a mature woman, the absence of a belly of an adolescent girl. The sexual appeal of an adolescent girl is to an adolescent boy (of whatever age), not to a man.

The fact is that the sucked-in belly cuts off all sexual feelings in the pelvis, those lovely melting sensations which transform sex from mere performance and release into an expression of love. What many women really feel about letting the belly out is that it is *too* sexual. Sloppy means loose and loose implies a loose woman. In Victorian days women wore corsets to contain their sexuality; they literally could not be regarded as loose women. While we have rejected the physical corset, we have adopted a psychological corset that is even more effective because we cannot take it off at will.

Many men also object to letting the belly out. They are afraid that they will develop a big "pot," which admittedly is unattractive. But when you look at people with a pot you find that the belly is not let out there either. It is tightly contracted and the muscles of the abdominal wall are drawn taut and spastic. There is a constricting band at the level of the navel or the pelvic crests. The pot protrudes above this constricting band, which functions like a dam preventing the downward flow of feeling and excitation. Energy in the form of fat piles up above the dam, producing the bulge so commonly seen in middle-aged men. In time, the tight abdominal muscles also tend to collapse, increasing the bulge.

The following figures show how the bulge develops. Figure 3 shows the natural belly-out posture of an adult person. In figure 4 a pot has developed as a result of damming the downward flow of excitation by tension in the lower abdominal wall. In figure 5 the pot has turned into a paunch as the upper abdominal muscles weakened and stretched under the pressure of the bulge.

If the dam can be broken, that is, if the band of tension can be released, the pot will slowly disappear. I have seen this happen to many men. It can only be done if the person can become aware of the constriction and tightness.

The surprising thing is that most people cannot let their bellies out. The holding in has become part of their way of being and cannot be overcome easily. When they try to let it out (lower abdomen), they find that it goes out only very slightly. Then as soon as their attention is directed elsewhere, it gets sucked in again. The same is true of the locked knees. One can keep them slightly bent when one is conscious of the knees, but they tend to lock up again when one isn't thinking of them. It takes a lot of practice to break these bad habits.

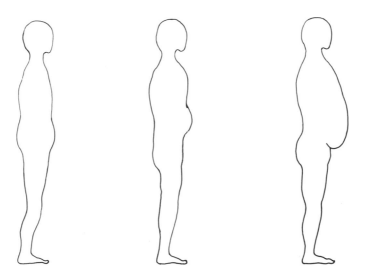

Fig. 3. Normal belly   Fig. 4. Pot belly   Fig. 5. Paunch

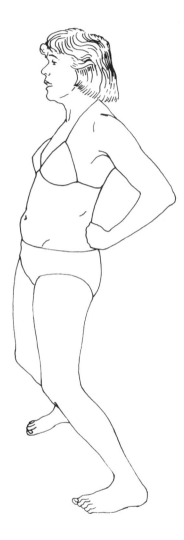

Fig. 6. Bow or arch

Exercise 3 / Letting the belly out

Stand with your feet about 8″ apart and as straight as possible. Bend the knees slightly. Without raising the heels from the floor, lean forward so that the weight of your body is on the balls of the feet. Keep your body straight but not stiff (see figure 3). Now let the belly (lower abdomen) out as far as it will go. Breathe easily for one minute.

The purpose of this exercise is to enable you to sense the tensions in the lower part of the body.

• Can you let your belly out?
• Does it stay out or do you find that it gets pulled in again?
• Does this posture make you feel "sloppy" or "let down"?
• Do your legs feel shaky? Are you afraid they won't hold you up?
• Do your breathing movements extend into the lower abdomen? Are you breathing into your belly?

Exercise 4 / The bow or arch

This exercise is similar to the preceding one except that it puts the body under stress to open up the breathing more fully and to place more strain on the legs. If done correctly, it helps to release the tension in the belly that causes the pot. A similar exercise is done by practitioners of T'ai chi.*

Stand with feet about 18″ apart, toes slightly turned inward.

Now place both fists with the knuckles facing upward into the small of your back.

Bend both knees as much as you can without lifting the heels off the floor.

Arch backward over your fists, but make sure that your weight remains forward on the balls of your feet.

Breathe deeply into your belly.

• Do you feel any strain in your lower back? If you do, it indicates that there is considerable tension in this area of your body.

• Do you feel any pain or tension in the front of the thighs or above the knees? If your legs are relaxed, you should feel no strain except in the ankles and feet where the weight of the body is supported.

• Are your legs beginning to vibrate?

• Are you able to maintain a perfect arch? Is your ass pulled back or is it pushed forward? In either case, you have broken the bow and your energy and feeling will not flow fully into your feet.

Repeat Exercise 1 / Basic vibratory and grounding exercise

All exercises in which a person arches backward either in the bow position or over the bioenergetic stool (see chapter 10) are regularly followed by one in which the person bends forward. This not only relieves the stress and increases the flexibility of the body, it also promotes the discharge of the excitation built up in the preceding exercise. The vibrations in the legs is such a discharge.

Repeat the exercise described in the preceding chapter on p. 11. Bend forward and let the fingertips touch the floor without putting any weight on them. Start with bent knees, then *slowly* straighten your legs until you feel the vibrations start. Do not lock your knees backward as this immobilizes the legs.

Breathe easily and deeply.

Hold the position for about one minute.

---

*T'ai chi is a program of Chinese exercises that has been practiced for centuries in China. They aim at grounding the individual and giving him a sense of harmony with the universe. There is a similarity between these exercises and bioenergetics. The bioenergetic exercises focus upon and are designed to relieve specific problems.

The dynamics of the bow position are explained more fully in my book *Bioenergetics*.

• Do you feel the vibrations in your legs?

• Are they stronger than when you did the exercise earlier?

Let yourself come up to a standing position with your knees slightly bent. Relax as you did during the first exercise in this series, letting the belly out and breathing easily.

• Are your legs still vibrating?

• How do you sense your feet in relation to the floor? Do you feel more connected to the floor or more grounded, as we say?

• Are you more aware of your legs and feet? Do they feel more "there" for you?

Grounding is the key to bioenergetic work. If you are well grounded, your body will be naturally balanced, upright, and firm. Your energy will flow freely. You may even notice that your eyes are clearer and your vision better.

Grounding is closely related to breathing, as you may have observed while doing the exercises. The more you let down inside yourself, the deeper is your breathing. It is important, however, to become aware of your breathing pattern and to learn how you "hold" against free and full respiration. That will be the subject of the next chapter.

3

Breathing

Good breathing is essential to vibrant health. Through breathing we get the oxygen to keep our metabolic fires burning—and these provide us with the energy we need. More oxygen creates a hotter fire and produces more energy.

But most of us are aware of the importance of breathing. And bioenergetics doesn't concentrate heavily on breathing exercises. If we ask you to be conscious of your breathing, it is only to help you become able to breathe easily and deeply *naturally* without being conscious of it. Our focus is on helping you sense and release the tensions that prevent you from breathing naturally. Breathing is not normally something we should be that conscious of. An animal or a young child breathes correctly and needs neither instruction nor help to do this. Adults, however, tend to have disturbed breathing patterns because of chronic muscular tensions that distort and limit their breathing. These tensions are the result of emotional conflicts that have developed in the course of their growing up.

Breathing exercises help somewhat, but they do nothing to reduce the tensions and restore natural breathing patterns. One has to understand these natural breathing patterns and know why they become disturbed, and one has to learn how to release the tensions that disrupt the natural breathing pattern.

The pattern of relaxed breathing (when one is not in a state of great exertion or strong emotion) is downward and outward in inspiration (inhaling air). The diaphragm contracts and descends, allowing the lungs to expand downward as they inflate. This is the direction of least resistance for the expansion of the lungs. The abdomen enlarges through an outward movement of the wall to make room for the downward movement of the lungs. The contraction of the diaphragm also raises the lower ribs, which motion is assisted by the contraction of the intercostal muscles (those connecting one rib to another). The chest also expands outward in this process, but relaxed breathing is predominantly abdominal rather than thoracic (chest) breathing. In such breathing one takes in a maximum amount of air for the minimum effort.

Healthy breathing is a total body action; all the muscles of the body are involved to some degree. This is especially true of the deep pelvic muscles that rotate the pelvis slightly backward and downward during inspiration to enlarge the belly and then rotate it forward and upward to decrease the abdominal cavity during expiration. This forward movement of the pelvis is aided by a contraction of the abdominal muscles. Expiration, however, is mostly a passive process best exemplified by a sigh.

These pelvic movements are illustrated in figure 7, p. 27, which accompanies an exercise in breathing. In that exercise you will be asked to rotate the pelvis to sense the effect of this movement upon breathing.

One should think of respiratory movements as waves. The inspiratory wave starts deep in the pelvis and flows upward to the mouth. As it passes upward, the large cavities of the body expand to suck in the air. These cavities include the abdomen, the thorax, the throat, and the mouth. The throat is especially important: unless it expands on inspiration, one cannot take a deep breath. In too many persons, however, it is severely constricted or contracted to choke off feelings, particularly feelings for crying and screaming. It is very common in bioenergetic work that after a person has had a good cry his breathing becomes deeper and easier. Letting oneself sob releases the tension in the throat and also opens the belly.

The expiratory wave starts at the mouth and flows downward. When it reaches the pelvis, that structure moves slightly forward as mentioned earlier. Expiration induces a relaxation of the whole body. You let go of the air in your lungs, and in the process you let go of any holding. People who are afraid to let go have difficulty in breathing out fully. Even after a forced expiration their chests remain somewhat inflated.

The inflated chest is a defense against the feeling of panic, which is related to the fear of not being able to get enough air. When a person with this condition lets the air out fully, he experiences a momentary panic to which he reacts by taking in air and inflating the chest again. The inflated chest holds a large reserve of air as a measure of security. The person is afraid to let go of this illusory security. On the other hand, people who are afraid to reach out to the world actively have difficulty in breathing in. They may become terrified if they open their throats widely to take in a deep breath. It is a good rule in doing these exercises, therefore, not to force your breathing. See what you can accomplish without straining yourself.

There is another breathing pattern that comes into play when the need for oxygen becomes urgent, for example, as in very strenuous activity. Here, the muscles of the thorax are mobilized, and the whole chest becomes actively engaged in the respiratory movements. This pattern is superimposed upon the first so that the person is now breathing abdominally and thoracically, as a result of which his breathing is deeper and *fuller*. In both patterns the total body wall seemingly moves as one piece, although one can see the respiratory waves flow upward and downward.

These patterns are disturbed when one part of the body moves in opposition to the other. In some people when the chest expands in inspiration, the belly is sucked in. This produces a severe disturbance, for despite the considerable effort involved in expanding the rigid chest, you obtain little air since the downward movement of the lungs is blocked. Instead of breathing in and out, you are now breathing up and down with little expansion of the body cavities. More commonly, the respiratory movements are limited to the midriff area with little involvement of abdomen or chest. This is typical shallow breathing. At times there is some abdominal movement in respiration, but the chest remains rigidly held.

In the preceding chapter we attributed the held-in belly to sexual inhibition. But the belly is also contracted and held to suppress feelings of sadness. We suck in the belly to control our tears and sobs. If we let it go, we are liable to have a real belly cry. But then we also open the door to the possibility of a real belly laugh. Whether we cry or laugh, it is in the belly that we experience life on the gut level. Here is where life is conceived and carried. Here is where our deepest desires have their inception. If you are intent on suppressing your feelings, keep your belly tight. But then you must accept the fact that you will not be a vibrantly alive person. And if you complain of an inner emptiness, you should realize that you are blocking your own fullness of being.

Tears are like rain from heaven and a good cry like a rainstorm that clears the air. Crying is the basic mode of releasing tension, as anyone can see by watching an infant break into crying when his frustrations create an unsupportable tension. No one need ever be ashamed of crying, for we are all infants at heart. Considering the pain that most of us have experienced in our lives and the frustrations that we are continually subject to, we all have good reason to cry. Crying is so therapeutic that if a depressed person can cry, his depression will lift immediately.

Breathing is also connected with the voice. To make a sound you must move the air through the larynx. And as long as you make a sound you can be sure of breathing. Unfortunately, many people are inhibited in making a loud sound. Some are victims of the adage that children should be seen but not heard. Others choked off their crying and screaming because these expressions met with a hostile response from their parents. Choking off these sounds produces a severe constriction in the throat, which seriously limits breathing. For these reasons persons in bioenergetic therapy and in the exercise classes are often encouraged to vocalize or make a sustained sound while doing the exercises or breathing. A clear sound resonating in the body causes an inner vibration similar to the vibrations we induce in the musculature.

There are two other commandments in bioenergetic work. Do not hold your breath. Let yourself breathe. While we do not want you to force the breathing, we do want you to be aware when you are not breathing. If you become aware that you are holding your breath, give a sigh. The other commandment is to make a sound. Let yourself be heard. If you make a sigh, make it audible. Many people have developed problems because as children they were strictly admonished to be quiet. This denial of their right to use their voice may have led to the feeling that they don't have a voice in their own affairs.

Now we ask you to do a few simple breathing exercises to learn about your own breathing pattern. In doing these exercises, allow yourself to moan or groan whenever you feel that they are stressful or painful. You will find that making a sound diminishes both the stress and the pain.

Exercise 5 / Belly breathing

Lie on a rug on the floor. Bend your knees. Your feet should be flat on the ground about 15″ apart, toes slightly turned out. Bring your head back as far as it will comfortably go to extend the throat. Place both hands on the belly above the pubic bones or area so you can feel the abdominal movements. Breathe easily with your belly through an open mouth for about a minute.

• Did you feel your belly rising with each inhalation and falling with each exhalation?

• Did your chest move in harmony with your abdomen or was it rigid? Try to let it follow the movement of your belly.

Fig. 7. Belly breathing

• Did you feel any tightness in your throat?

Exercise 5-A / Variation—Rocking the pelvis

Now rock the pelvis slightly backward with each inspiration and bring it forward with the expiration as shown in figures 8 and 9. Do this breathing for about a minute.

• Can you sense that the pelvic movements increase the depth of your breathing and the amplitude of the abdominal movements?

You may find that this breathing produces tingling sensations in your hands or in other parts of your body known as paresthesias. You may also find your hands becoming cramped. Both symptoms are a sign of hyperventilation. If they become strong, just stop the exercises and they will fade away. They are not dangerous but your hands may develop a spasm, which is sometimes painful.

Hyperventilation is overbreathing. You have taken in and expelled more air than you normally do in a condition of rest. Bioenergetically we would say that your body is overcharged. After you have been doing these exercises for a time, you will

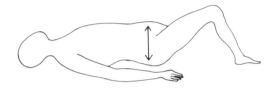

Fig. 8. Belly breathing—inspiration (belly out, pelvis back)

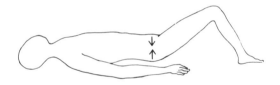

Fig. 9. Belly breathing—expiration (pelvis forward, belly in)

observe that the same amount of breathing will no longer result in any symptoms. As your body has become habituated to a deeper level of breathing, you are no longer overcharged. The paresthesias will also disappear if any emotion breaks through—should you start crying, the tingling will immediately stop for you will have discharged the excitation.

Exercise 5-B / Variation— Breathing out

This variation will help you sense how fully you can let the air out of your lungs. Letting the air out is equivalent to "letting go."

Lying in the same position as in exercise 5, make a moderately loud sound such as "ah" and keep the sound going as long as you can do it without forcing. When it stops, take an easy breath and start again. Do this exercise four or five times and observe whether you can hold the sound longer each time.

Be sure not to force the sound. Forcing the sound or the breathing only tightens your throat and produces tension.

You may find that your voice begins to tremble toward the end of the sound. You may begin to sob. This is OK. Let go and have a good cry if it comes easily. Crying will do more for your breathing than any other exercise.

Exercise 6 / Breathing and vibrating

Here is another exercise that will help you breathe more spontaneously.

While you are lying on the floor, put your legs up in the air. Your knees should be slightly flexed. Bend your ankles and push upward with your heels.

Your legs should begin to vibrate.

Keep your legs vibrating with the heels thrust upward.

Notice that your breathing is becoming deeper. See figure 10 on the next page.

• Did your belly feel tight? Could you let it out? You can do this by keeping your buttocks against the floor.
• Notice, too, that your breathing was sparked by the vibrations of your legs.
• After you do this exercise for a minute, return your feet to a resting position on the floor. How is your breathing now?

Following the breathing exercises, notice how relaxed you have become. Do these three simple breathing exercises anytime you feel the need to let down and relax. They will take five minutes at most.

Fig. 10. Legs-up vibration

The importance of breathing cannot be overstressed. The breath is so closely connected with life that it has been identified with the vital spirit. According to the Bible, God, in creating Adam, took a lump of clay and breathed life into it. The Greeks use the same word, *pneuma,* for breath and spirit. In the teachings of Yoga the vital force that animates all life is called *prana.* The main source of *prana* for the human being is the air. By breathing we absorb *prana* into our bodies. The Yoga disciple does special exercises to control and regulate his breathing so he can store up *prana.* These exercises are called *pranayama* and are the basis of the system of Hatha Yoga. " 'For breath is life' says an old Sanskrit proverb, 'and if you breathe well you will live long on earth.' "*

However, there is a difference between the breathing in Yoga and in bioenergetics. Our aim is not to give you a religious or mystical experience but to help you be more alive and more aware of yourself and your surroundings. Our focus therefore is upon natural respiration, breathing that is easy, deep, and spontaneous. It is not a matter of *making* yourself breathe but of *letting* yourself breathe. Every disturbance of natural breathing is due to unconscious holding patterns or muscular tensions. You may not breathe fully for fear that you might erupt in a scream. If that is your problem, find a private place and let it out. A car on a highway is an excellent place to scream; no one can ever hear it. Screaming is an old-fashioned release technique Victorian ladies knew very well. It still works wonders.

*S. Yesudian and E. Haich, *Yoga and Health* (New York: Harper & Bros., 1953), p. 79.

4

Sexuality

Bioenergetics is rooted in the principle that since the organism is a unity, health is also a unitary concept. This means that there is an identity between physical health and mental health, between emotional health and sexual health. The unity of the organism can be depicted as a circle. Each aspect of its health is related to the other and reflects its total health.

A break in the unity of the circle at any point will disrupt the organism's integrity and affect its health on every level. Thus, for example, sexual anxieties and problems will seriously affect the physical, emotional, and mental health of a person. The same would be true if there were a disturbance of any other aspect. The effect is always total.

To understand this concept, we must think of health in positive terms. Physical health is seen as more than the absence of debilitating symptoms. It is manifested in a body that is beautiful and graceful, that is vibrantly alive, not just free from disease. Such a body indicates the operation of a calm and clear mind in which there are no repressed conflicts. Similarly, emotional health is positively defined: it is having full possession of one's faculties and the full range of one's feelings. Naturally, this definition includes the ability to fully feel and express one's sexuality and the capacity to experience the joy of this expression. That would be our definition of sexual health. Basically, being vibrantly alive could be equated with the capacity for pleasure and joy in living.

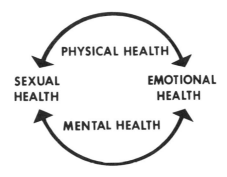

Fig. 11. Unity of the organism

In an earlier chapter we mentioned that one of the reasons the belly is held in is to control and limit sexual feeling. This holding also seriously restricts respiration and reduces the sense of being grounded. If one seeks to be vibrantly alive, the pelvis must be freed and the flow of sexual feeling opened up. How a person holds his pelvis is therefore as important a subject of study as how he holds his head.

The most common pelvic disturbance is the tucked-in ass. Here the pelvis is pulled forward and the ass is held tight like a whipped dog with its tail tucked between its legs. As a result the lower back is fairly straight with the normal lumbar curve (shown in figure 13) eliminated. One effect of this postural attitude is to place great stress upon the lumbosacral area (the lower back). One can feel this stress by curling up in a hard, straight-back chair with the ass forward. The pressure is felt in the lower back and abdomen. This pressure is immediately released if the ass is pulled back and one sits up straight. On the other hand, one can curl up in an easy chair because the weight of the body is distributed along the whole back and the thighs. Most cases of lower back trouble that I have seen occurred in persons with tucked-in asses and straight lower backs. However, in any person whose pelvis is held immobile in either the forward or backward position there is a predisposition to lower back pain. Since both these postures in themselves create considerable tension in the muscles of the lower back, the person is in trouble when he comes under an additional stress, whether it is of an emotional or physical nature.

The following three figures illustrate the different points of stress according to the body's posture.

Figure 12 shows a healthy body alignment. Note that the weight of the body is

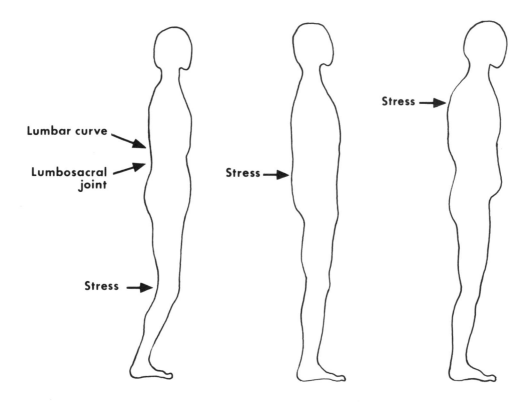

Fig. 12. Normal stress   **Fig. 13. Lower back stress**   **Fig. 14. Upper back stress**

34

carried forward on the balls of the feet. The body is balanced because the pelvis is pulled back slightly but held *loosely*. Note, too, that the knees are bent, which enables them to function as shock absorbers in any situation of stress. In this posture the stress of gravity and the stresses of life are transmitted through the vertebrae to the pelvis and through the pelvic bones to the hip joints. Since the legs are properly aligned with the body (which occurs when the toes are pointed straight and the knees centered over the feet), the weight of the body plus any additional stress passes through the legs to the feet and ground. This shift of stress to the legs can only take place when the pelvis is tilted backward.

In figure 13 the tucked-in ass and locked knees focus the stress upon the lower back, creating a predisposition to lower back trouble. Note that the weight of the body is over the heels. The body leans backward in a passive position.

In figure 14 the stress is carried in the upper back which causes a bulging of this area. The head is bowed forward, and the lower back shows an exaggerated curvature called a lordosis. It could be said of a person with this structure that he is carrying someone on his back.

What has the backward tilt of the pelvis to do with sex? When the pelvis is forward, it is in the discharge position. This means that any sexual feeling the person has will flow directly into the genitals, the organ of discharge. When the pelvis is held back but loose, it is in the position of charge. It can fill up with sexual feeling. We can draw an analogy with the hammer of a gun. In the backward position the hammer is cocked and ready to fire. In the forward position it is discharged. So it is with a person or an animal. When the animal holds its tail high, it is an expression of good spirits and excitement. A person in this state could be called "bright-eyed and bushy-tailed." He or she is, in other words, "cocky."

We are speaking here of sexual feelings, not just genital excitement. The belly or lower abdomen is the reservoir of sexual feelings. When the pelvis is held forward and the belly held in, this reservoir function is largely lost. Since the person cannot now "contain" sexual feeling, his only choices are either to "act out," that is, seek a sexual release wherever he or she can, or where this is impossible, to cut off the feeling. One does this by holding the breath and immobilizing the pelvis. As a result the person is not sexually alive and has to be "turned on." Such a person is not grounded in his sexual nature.

A sexually alive body is characterized by a free-swinging pelvis. Free-swinging means that it moves spontaneously and that it is not being pushed, bumped, or ground out (grinding the pelvis is done by persons whose pelvises do not move freely). We noted earlier that the pelvis moves spontaneously with each breath, forward on expiration, backward on inspiration. It also moves freely and easily with every step we take. One has only to observe the way native women in the Caribbean or South Sea Islands walk to see the beautiful sway of the hips that is part of their natural grace. The men also walk with a similar looseness of the hips, though it is less noticeable. In contrast, people in our more sophisticated cultures walk stiffly with tight asses.

These exercises will not free you from any sexual hang-ups you may have. That is a task for therapy. Repressed sexual memories dating back to childhood must be recovered and the subtle sexual relationships that exist between parents and children uncovered. But these exercises not only help with therapy, they are essential to therapy. It is not enough to free a person from the sexual anxieties in his mind; it is also necessary to free his body from tension and to restore the mobility of his pelvis. And this can only be done through a physical approach.

To be effective, the physical approach must involve the whole body. It should start with some vibratory activity in the legs. Sooner or later this will extend upward to include the pelvis. Next, it is important to develop the feeling of being grounded, since adult sexuality is related to the sense of independence and standing on one's own feet. Grounding gives a person the feeling of independence and maturity that makes his sexual expression a responsible activity of his total being. Finally, an individual's breathing has to be opened and deepened into the belly so that the pelvic movements are coordinated with the respiratory waves. This enables the total body to participate in the orgastic response.

Finally, it is most important not to tighten the ass. This is done by pulling up the pelvic floor and pulling in the anus. These tensions represent a fear of "letting go"; that if we did, we would defecate and make a mess. Originating in the early childhood training for excremental cleanliness, these tensions are now unconscious and block the full giving in to the sexual discharge. In the following exercises and all succeeding ones we ask you to try to drop the pelvic floor and push out your anus as if going to the toilet. You will not mess. The internal sphincter of the anus remains closed. This opens only when there is fecal matter to be evacuated. If you have any anxiety on this score, move your bowels first.

Exercise 7 / Hip rotation

At this point, you may want to try a simple exercise to test your sexual responsiveness and sense your pelvic tensions.

Stand with your feet about 12'' apart, straight and parallel, knees slightly bent, and the weight of your body on the balls of your feet. Shoulders should be down, chest soft, and belly out. Place hands on hips. In this position try to slowly rotate your hips in a circle from left to right. The movement should occur mainly in the pelvis and only minimally involve your upper torso and legs.

After half a dozen circles from left to right, reverse the direction and make the same number of circles from right to left. See figure 15 on the next page.

• Were you holding your breath? Try to keep breathing with the movement.

• Did your belly tighten up? If so, you were cutting off sexual feeling. Try to let your abdomen stay soft.

• Could you keep your anus open and your pelvic floor relaxed? Did you forget these parts?

• Were you able to keep your knees bent?

• Were you able to keep your weight on your feet or were you pulling off the ground?

• Did you feel any pain or tension in your lower back or thighs? These areas are very tight in most people.

We do not wish to imply that if you do this exercise easily you are free from any sexual tensions or problems. The reverse, however, is true. If you cannot do this exercise easily, you do have a problem. In this, as in the other exercises, the important criterion is whether the person is grounded. If one is not grounded, the pelvic swings lack an emotional tone. To appreciate this fact, consider what happens to a guitar string that is free at one end. If it is plucked it will move, but the movement does not produce a musical sound. This happens only when the string is attached at both ends and under the proper tension.

Exercise 8 / Arching the back and rocking the pelvis

Another sexual exercise may bring your pelvis and lower back tensions more fully to consciousness.

Lie on the floor with knees bent so that your feet are parallel and flat on the floor. Arch your lower back and press your buttocks against the floor. As you do this, breathe in, letting the belly out as much as possible. Then breathe out and let the pelvis rotate forward, pressing down lightly on the feet to give it a slight lift. Next breathe in again and rock the pelvis backward, arching and pressing the buttocks against the floor. Do this exercise for about fifteen to twenty breaths. Your breathing should be slow. See figure 16.

• When your pelvis came forward,

Fig. 15. Swinging the hips

38

did your belly tighten? If it did, you were lifting the pelvis with your abdominal muscles instead of with your feet and thigh muscles.

• Were you squeezing your ass as the pelvis came forward? If so, you cut off sensation in the buttocks. Try to keep the ass soft.

• Did you lose the feeling of your feet on the floor at all times? If your feet lose contact with the floor, your pelvis will not be free in its movements.

• Could you feel the breathing movements in your pelvis? The coordination of the respiratory and pelvic movements is not easy to achieve.

• Did you feel embarrassed or ashamed of making these sexual movements? This is a good opportunity to explore your attitude to sex. For all our current sexual sophistication, most people have a deep sense of shame about acknowledging their sexuality in soft, undulating pelvic movement.

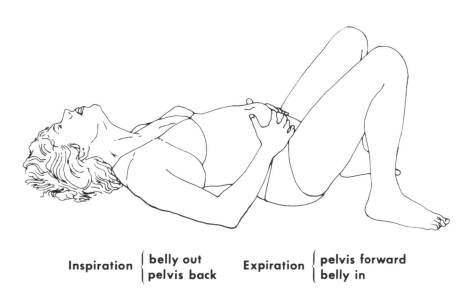

Inspiration { belly out / pelvis back }   Expiration { pelvis forward / belly in }

Fig. 16. Pelvic movement and breathing

39

Because of the prevalence of underlying passivity and tension in the lower part of the body, it is very common for people to pull or push the pelvis forward rather than let it swing freely on the hip joints. Pulling the pelvis forward is done by contracting the abdominal muscles; pushing it forward is done by contracting the ass. Both of these actions reduce sexual sensation and block the involuntary pelvic movement, which should occur at the climax of the sexual act.

If the pelvis begins to move spontaneously in the course of this exercise, you will develop some lovely feelings there. You will not, however, have an orgasm. The genital organs will not become excited unless you deliberately fantasize a sexual encounter. However, this is not recommended since it will focus your attention upon the genitals and distract you from the perception of what is going on in your pelvis. When this exercise is done in a therapy session or an exercise class, genital excitement very rarely develops. You may wonder what to do, however, should you become genitally excited when doing this exercise at home. There can be no valid objection if, in the privacy of your home, you should want to masturbate. It is a normal activity that helps a person accept and find pleasure in his or her own sexuality. But the purpose of these exercises is not to stimulate genital feelings.

In bioenergetics we believe that feelings can be consciously contained or expressed depending upon the circumstances of the situation. Your genital excitement will diminish and disappear when you begin another exercise. Feelings do *not* have to be expressed or acted out. We are interested in gaining the ability to express our feelings, but when or how well we will do so depends upon a conscious determination of the appropriateness of our actions to the situation. The development of *conscious* control over your feelings is an important factor in self-possession.

An individual's sense of self is anchored in his or her sexuality. Sexual anxiety, guilt, or insecurity weakens this anchorage and undermines the strength of one's ego. To build one's ego in a positive way it is necessary to work through one's sexual problems. But it is equally necessary to work directly with ego problems involved in such ego functions as self-possession and self-expression.

5

Self-Possession and Self-Expression

Bioenergetics, like other therapies, aims to help a person develop a better sense of self, that is, to be more of a person. The self, however, is not an abstract quality; rather, it is the entirety of one's functioning. The self cannot be divorced from self-expression, for it is in our expressive activities that we perceive the self. Nevertheless, contrary to what some people think, it is not necessary to go about consciously trying to express one's self. The greater and more important part of our self-expression is unconscious. A gracefulness of manner, the sparkle in one's eyes, the tone of voice, an overall aliveness and vibrancy express who we are more than words or actions. However, these are not qualities one can deliberately cultivate. They are the manifestations of emotional and physical health.

If a person is blocked in his ability to express feelings, it will deaden his body and reduce his vitality. In therapy, ways must be found to help a person become free to be able to express feeling. It is very common to see people who are unable to cry, who cannot become angry, who are afraid to show fear, who cannot reach out in appeal, who dare not protest. Some people can cry easily but cannot show anger; in others it is the reverse.

Bioenergetic exercises provide an opportunity for people to practice and become familiar with the expression of feeling in a controlled setting. This is not an encounter procedure, for the expression of feeling is not directed at any other person.

But just as a person is encouraged to express his feelings in an appropriate exercise, he is also helped to gain conscious control over their expression. The purpose of this control is not to inhibit or limit the feeling, but to make its expression effective, economical, and appropriate. A hysterical outburst can be considered an expression of feeling, but it is often wasteful of energy and relatively ineffectual. It is not really a form of self-expression because it erupts against the conscious intention of the person. It is not ego directed. It denotes a lack of self-possession and often results in a diminution of the self.

Self-possession denotes the ability to respond appropriately to a situation. One does not fire a cannon at a rabbit, and it is equally inappropriate to go into a rage over a minor irritation. Timing is of equal importance. When to act and when to speak are as critical as what one does or says. Some people react too quickly; they are impulsive and lack the conscious restraints that characterize a person with self-possession. Others react too slowly, often long after the situation has passed. Poise implies good timing.

We all admire people who have poise: they are ready for action and in command of themselves. Poise is therefore a synonym for self-possession, the fine coordination of feeling and action, of the involuntary or spontaneous movements and the voluntary or deliberate movements, of the ego and the body.

Poise is developed by increasing one's coordination in all expressive actions. When one makes a movement, it should involve the total body, regardless of how small or large the movement is. If any part of the body does not partake to some degree in the movement, the person is uncoordinated. Thus, he will sense a lack of poise.

Let us suggest an exercise to help you assess your degree of self-expression and self-possession. It is kicking, which expresses the idea of protest. Further, it involves the lower part of the body, which is passive in so many people. If you have difficulty in identifying with this action, think of some injustice you have experienced. Everyone in this culture has something to kick about.

Exercise 9 / Kicking

Lie on a bed, preferably one with a foam mattress and no footboard. Or you can use a foam rubber mat about 5'' thick on the floor. Extend both legs. Keeping them loose, with the knees extended but not rigid, kick up and down rhythmically. Your ankles should also be loose, and the blow should land on the heel and calf. Do the kicking easily at first, then gradually increase the strength and tempo of your movements. Finally, holding on to the sides of the mattress or mat, kick away with all your strength and as fast as you can.

This kicking action is like the snap of a whip. If you are coordinated, your head will bob up and down with each kick. If you are afraid to let go (of your head), your movements will be mechanical. See figure 17 on the next page.

• Did you stop abruptly or did you let the movements peter out? Stopping abruptly is like jamming on the brakes and indicates a fear of letting the movement carry through to its natural conclusion.

• Did your knees bend so that the blows were hitting only on the heel? That kind of movement results from excessive tension in the hamstring muscles of the legs.

• Of course, you were out of breath at the end. It is a rather violent exercise. Did you become panicked at the loss of breath? Were you dizzy at the end? Both the panic and the dizziness pass as you breathe easily again.

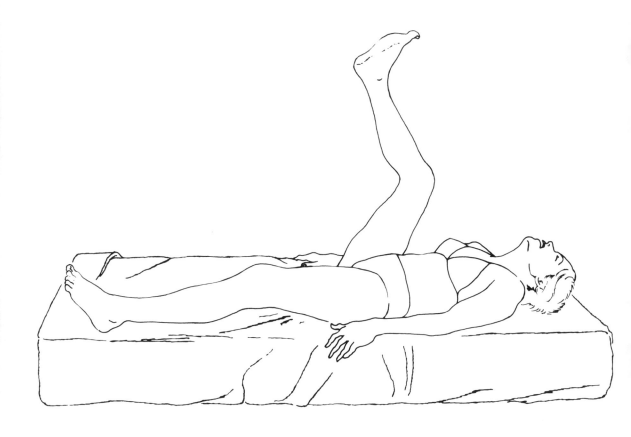

Fig. 17. Kicking

44

Exercise 10 / Saying "no" while kicking

To make the exercise stronger, try saying "no" while you are kicking. The "no" should be sustained as long as possible and repeated several times during the exercise. Now you are kicking a strong protest.

• Was your voice strong and full or weak and hesitant?
• Using your voice requires greater coordination. Was this exercise more difficult for you?
• Were you frightened by the sound of your voice?
• When was the last time that you kicked like that?

If such an exercise upsets you, do not repeat it immediately. The exercise is too strong for you. Build up toward it and develop coordination and expression by working more easily and steadily. Our recommended procedure is to do this exercise regularly but easily, emphasizing the rhythmicity and looseness of the movement.

Exercise 11 / Building up your kicking power

Do the same exercise without the use of the voice and with a moderate intensity. Give the bed fifty kicks and see how smooth and rhythmical your actions are.

If your legs tire or you run out of breath before you do fifty kicks (counting each leg as one kick), start with twenty-five or thirty.

Each day try to add five to ten more kicks to the exercise until you reach one hundred.

When you feel at ease with 100 kicks at a time, try to move it up to 150. Then try for 200. When you are able to do 200 kicks at a time, you are building both sustaining power and coordination.

As you continue to do this exercise, you will find that it becomes easier and that you are gaining more freedom in the lower part of your body. You are repossessing parts of your body that were formerly unavailable.

In the latter part of this book, in the chapter on expressive exercises, we will describe other expressive actions such as hitting, reaching, looking, and so on. There are other kicking actions, too, which use the body in different ways and which can help promote one's coordination and self-possession. Before you go on to these, be sure to read the words of caution and counsel in chapter 7.

6

Being in Touch

One of the characteristics of aliveness is the quality of being in touch. You might ask, in touch with what? In touch with everything within the range of one's sense perceptions. Being in touch is being aware of what is going on within and about you. It is quite different from knowing, which is an intellectual rather than a perceptual activity.

All sensing starts with a sensing of the self, that is, of one's own body. Through this, one perceives what is going on in the environment since the environment impinges upon our bodies and our senses. The more alive one is, the sharper one's sensing and the keener one's perceptions. Haven't you noticed how much clearer and more distinct everything is when you are feeling very good? Similarly, when you are depressed everything looks gray and indistinct. The way to increased sensing is through increased aliveness, but this works in reverse, too. If one's sensing is limited or narrow, it diminishes one's aliveness.

One of the main purposes of these bioenergetic exercises is to help you sense or get in touch with your body. This is necessary because too many people live in their heads, with very little consciousness of what is going on below their necks. They are not aware when they hold their breath or whether their breathing is shallow or deep. Most people do not sense their legs and feet. They know that they are there, but they use them merely as mechanical supports. Sensing is not a mechanical function. An automobile may run very well, but it senses nothing. Sensing is a function of feeling.

Exercise 12 / Backward stretch

Here is a simple exercise that will help you sense a part of your body you are not normally aware of. Let us assume that you are sitting in a chair while reading this book. Raise your arms and arch backward over the back of the chair. Make a good stretch and hold it for about thirty seconds. While doing so, breathe easily and deeply through your mouth. See fig. 18 on next page.

• Did you feel your back pressing against the chair? Did you sense whether your back was tight or relaxed? Was it painful? Could you breathe easily in this position?

• When you stretched your arms backward, did you feel any tension in your shoulders?

• After you returned to your usual sitting position, were you aware that you tended to hunch forward? You may have sensed a need to stretch backward again to overcome this common tendency to hunch forward. Do this exercise again and sense how much easier it is the second time. Stretching the muscles of the back has relaxed them somewhat.

The importance of being in touch with the back of the body cannot be overstressed. Without a feeling of the back, it is very difficult to "back up" one's position. It is not enough to have a backbone (anatomically we all have one); a person has to feel his backbone. He has to sense if it is too rigid and unyielding or too soft and pliable. If it is too rigid, one cannot easily back down and yield in situations where such a response would be appropriate. If it is too soft, it will not provide enough rigidity to enable an individual to hold his position in the face of stress. He will fold up or yield too easily. Excessive rigidity is due to chronic tension in the long muscles of the back. Excessive flexibility is caused by a lack of

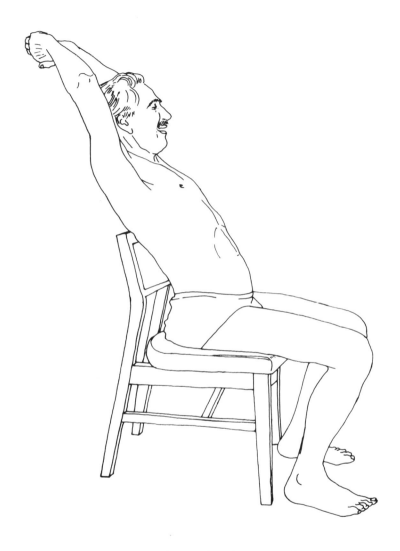

Fig. 18. Over a chair leaning back

48

tone in these muscles and spasticity in the small muscles joining the vertebrae. In both cases the back is not fully alive and cannot provide the aggressive drive one needs in life. The rigid person "backs up," that is, moves backward in a confrontation while the other type fails to stand up to the situation.

In the chapter on sexuality we showed how stress is supported by the legs and the back of the body. In figures 13 and 14 we saw how unusual stresses can produce a bowing of the upper back or a flattening of the lower back. Most people suffer from tensions in both of these areas and complain of discomfort or pain there. To help people release these tensions so that the body can be properly aligned, we use a number of exercises involving a bioenergetic stool. The stool and the exercises will be described in chapter 10. The idea for the stool was taken from the exercise you have just done.

To return to being in touch: since most people are out of touch with their bodies and rarely use them actively except in a mechanical way, the bioenergetic exercises seem strange and stressful at first. The positions appear unnatural; you may feel awkward and there may be some pain. However, you will begin to feel and sense your body differently. After a time, the realization will come that you have been out of touch with a large part of your body.

Getting in touch is a process of sensing the tightness and tensions that block the flow of excitation and feeling. Only by sensing a tension can one release it. Every tension is a chronic muscular contraction or spasticity. These contractions are in the large outer voluntary muscles and also in the inner small involuntary muscles of the trachea and bronchi, the intestinal canal, and the vascular system.

It should be said here that there is no nervous tension that is not related to chronic muscular spasticities and contractions. Many people are aware of being tense in a generalized way. They call it nervous tension because they are out of touch with the state of muscular tensions in their bodies. They do not sense the constriction that may develop in the throat, the tightness in the back of the neck and in the shoulder girdle, the spasticity of the diaphragm, or the knots in their leg muscles. Lacking these perceptions, they cannot release the muscular tension and are forced to resort to pills to reduce their state of nervousness. It is much better, though not easier, to work directly with the muscular tension to bring about a state of relaxation.

If one senses the tension, relaxation is achieved by mobilizing the contracted muscles in a stretch or expressive movement. Slowly stretching a contracted muscle will often cause it to "let go." The muscle will go into a tremor or vibration like a spring after the tension has been released. Expressive movements like kicking and hitting serve the same purpose by mobilizing the contracted musculature. An emotional release like crying will often relax the inner tensions as well.

Relaxation represents an expanded state of the organism in contrast with tension, which is the contracted condition. Relaxation therefore requires energy, and it can only be realized if the breathing is opened up along with doing the exercise. One senses the relaxed state in any part of the body by its increased warmth and better color, as more blood begins to reach the area. Contracted parts are cold and feel relatively lifeless to the touch. We describe them as "dead" areas, meaning simply that they are parts of the body with which the person is out of touch.

The process of getting in touch with the body is never completed. As you continue to do these exercises, you will make a deeper contact with your body, you will feel many parts of the body differently, you will develop new patterns of posture and movement. Your self-possession and self-expression will increase progressively.

Being in touch is not a state of perfection but of aliveness. Regardless of how long a person works with his body, there are always some tensions that persist. This is not a reason for discouragement about the exercises. It means that one has to do them regularly if one wants to stay in touch with the body. We must recognize that we do not live in a body-oriented culture as primitive people do. If anything, our culture is anti-body and, to that extent, anti-life. Machines do much of the work that we formerly did with our bodies, and while that makes life easier, it does not necessarily make it more alive or enjoyable. As machines have improved, the result has been a speeding up of the tempo of our lives. We move faster but have less time. In fact, as the pace quickens, it often leaves us no time even to breathe. Add to this the enormous social and competitive pressures of our culture, and it becomes quite clear that unless we counter these forces with a positive program of bodily activity, we cannot hope to maintain the feeling for the life of the body that is essential to vibrant health.

7

Counsel and Caution

These bioenergetic exercises are not designed to take the place of therapy, though they do have therapeutic value. Someone with a serious emotional or personality problem should not attempt to work it out by himself through these exercises because doing them for that purpose may lead to feelings that are more than he can handle alone. In such a situation competent professional help should be sought. However, these exercises can be done beneficially for a more general purpose, that is, not as therapy, but to get in touch with the body, to increase one's energy, and to feel more alive.

It may also happen that a person who is unaware of the depth or seriousness of his problems might enthusiastically embark upon these exercises and find that the new sensations and feelings which develop in his body leave him disturbed and confused. Here, again, the best advice is to seek professional counsel.

The body work will inevitably lead to sensing or getting in touch with suppressed feelings. As the body becomes more alive, you feel more. Feeling is the perception of internal movement, and it is the goal of these exercises to increase a person's capacity for movement and feeling. Thus, as the body begins to vibrate, the vibrations may increase and change spontaneously into the more intense convulsive movements of sobbing. The sobbing may be experienced simply as a relief, or it may be accompanied by a feeling of sadness without the person knowing why he is sad. Most of us have suppressed our sadness and crying in order to present a smiling face to the world. We were taught that no one likes to see a sad face. "If you cry, you cry alone" is a familiar motto. Now, as the body becomes alive, the mask breaks down and the sadness and crying erupt to the surface.

Should this happen, can you accept the feeling? If you can, our advice is to go with the feeling since feeling is the life of the body. But it is not just sadness that may break through. Fear and anger may also begin to come out. Remember that these feelings are not provoked by the exercises, merely evoked by them. These feelings have been suppressed by chronic muscular tensions and by deadening the body. Again, the question is: can you accept the feeling realizing that it has reference to a past situation? You need simply say, "Yes, I am frightened" or "I sense that I am angry." If you can stay with the feeling or contain it, you will benefit. You can also discharge the feeling by expressing it if you can handle the expression. One way of discharging fear in bioenergetics is by screaming, of discharging anger by hitting a bed or twisting a towel. These will be explained in more detail later in the book.

A problem arises during the exercises only if you are threatened or overwhelmed by your feelings. In that case, stop the exercise immediately and allow the feeling to subside. Nothing can be gained, and it can be dangerous to try to overcome an anxiety associated with feelings you cannot deal with. Here, as we said earlier, it is necessary to get some professional help. But if you can stay with the body awareness and keep your feelings under conscious control, you will become progressively more able to accept your feelings, contain them, and express them appropriately.

If you have any physical disability or illness, you should consult your physician before undertaking any program of exercises, including this one. The exercises by themselves are in no way dangerous or harmful to the body even in the case of illness, but they are *then* done with the approval of a medical doctor. We have used some of them on persons with medical problems to good effect. In those cases, however, the exercises were carefully gauged to ensure that the stress was not too much for the patients. The real danger is that you may "push" the exercise program beyond your ability to tolerate the stress, thinking that you have to release the tension.

This caveat is especially true for people with lower back problems, which are primarily the result of muscular tension. In some persons this problem may be complicated by an arthritis of the lower back or a herniation of a disc with pressure on the nerve roots. In neither condition are the exercises contraindicated, but they must be done with a sensitive awareness of the body. Many people have been helped to clear up a lower back condition through these exercises, but the procedure used never involved pushing or forcing.

It is a fundamental bioenergetic principle that no one can force a tension to release. The use of willpower creates tension rather than releases it. The body can be extended to the point of pain, thereby sensing the tension, but release can only occur by a "letting go" or "letting down." To let go, you have to sense or feel (1) that you are holding, (2) what you are holding against, and (3) why you are holding. If you can sense these things as you get in touch with your body, the "letting go" will happen by itself.

It is axiomatic in medicine that the body heals itself in most conditions. This should be true of tension states as well. If it doesn't happen, it is because we don't trust the body enough to "let go to it." This means giving in or going with whatever happens spontaneously in the body. We have been taught to control our bodies as if they were wild and dangerous animals. But this very control, once it becomes unconscious or ingrained, creates the tensions from which we suffer. It is not, therefore, a matter of doing more, but of doing less. Through these exercises we hope that you will learn what you are doing to keep your body in control, that is, tensing your muscles to keep your body rigid and relatively deadened. We don't ask you to *make* it come alive; we ask you to *let* it come alive.

Our most important word of counsel is not to regard the exercises as a performance. How long you can maintain a stress position means very little. Remember that steel can take more stress than any living being. If you can take only very little stress, it doesn't make you an inferior person. You can build your tolerance to stress through the exercises, not as a matter of will, but of tissue strength. That means more energy and more aliveness in your body. In doing the exercises, the main thing is feeling through movement and vibration. Focus on sensation. How well do you feel your feet, legs, pelvis and backside, belly, chest, shoulders, head and neck? Do you feel your back? And finally, do you feel your heart? If you can feel into your heart, you have reached the core of your being.

At first you may find the exercises not pleasurable but painful. You will be surprised when in time they do become pleasurable and really make you feel good. The pain is a reflection of the degree of tension in your body. As the tension is released, you begin to experience pleasure.

Don't do these exercises as if going through a drill. You are not your own drill sergeant. Do them slowly, taking time out to breathe and to feel. And since you are aiming for relaxation, do them in an easy, relaxed way. Don't be compulsive. If you miss a day or a week or even a month, no one will grade you or punish you. It's all right to lapse and stop; you can always start again.

How much time you devote to the exercises is a matter of personal inclination. Our exercise classes are about an hour. At home, people spend as little as five minutes up to as much as an hour. It is easier and more pleasurable to do exercises in a group, but it is more convenient to do them alone at home. In either case, there is no regular routine. We never do all the exercises at any one time but choose those that are most indicated by our need.

Probably the best time to do any exercise is in the morning after washing up or a hot shower. This generally sets you up for the day. Feeling more alive and more energized, your day will go smoother. A shower before exercising loosens your body after the night's sleep. However, any time is good, except after a substantial meal. If you are an overeater, you may find that the exercises will reduce your need for food. They will provide you with more energy and so obviate the desire for food.

Sometimes doing a few of the exercises at night, especially the deep breathing, will enable you to fall asleep more easily. Falling asleep involves getting out of the head and into the body. If you are overexcited and your mind is racing along, it is difficult to "let go" to sleep. Through the exercises you can get back into your body, thereby facilitating the descent into sleep. If you do too many of the exercises before going to bed, however, you may become too charged and too aware to fall asleep easily. You are now ready to undertake these exercises in a systematic way. This means that one shouldn't plunge into the sexual exercises without doing a preliminary warm-up and some grounding work. It does not mean that one has to do all of these exercises in the order given in this book. If you are new to these exercises, take time to familiarize yourself with a few at a time. Get the feeling of them and sense what they do for you, then try some others. When you are familiar with all the exercises in part II, choose those to do regularly that you find most beneficial. In part III there are suggestions for doing the exercises at home and for conducting a bioenergetic exercise class.

PART II

The Exercises

8

The Standard Exercises

The standard exercises follow a definite order, reflected in the chapter's section headings, which is to work with the body from the ground upward. No structure is stronger than its foundation: in the case of an adult, this is the legs and feet. The more feeling a person has here, the more contact he has with them and the ground he stands on, the more secure will be his foundation as a person. This work is especially important for people living in a culture such as ours, which orients them toward the head rather than the ground.

We will start with a short series of exercises designed to focus a person's awareness on his body and to orient him toward a way of standing that makes him conscious of his legs, feet, and the ground beneath him. This will be followed by a number of warming-up exercises, any one of which will help prepare the body for the more strenuous exercises that follow.

Considerable time should be spent on the grounding exercises, that is, on the work with the legs and feet, because of their critical importance to the overall work with the body. As we move upward, the next area to command our attention is the pelvis and hips. Several exercises are offered to help mobilize and loosen the muscles in this part of the body. We then focus on the arms, hands, and shoulders and end with a series of exercises designed to free some of the tensions in the area of the head and neck.

As one works upward with the body, it is essential not to ignore the lower part of the body. The awareness gathered through the grounding exercises is to be carried into the work with every other part of the body. In this way the whole body is gradually involved in every movement. The person moves as a whole, as a unit, and every movement, large or small, starts from the ground.

The standard exercises also include movements and maneuvers performed in the sitting and lying-down positions. A special section is devoted to the exercises most appropriate to each of these positions.

Focusing and Orienting Exercises

These exercises should orient you to your correct body alignment and help you focus on some areas of your body that usually need attention. In a class exercise 13 serves to bring everyone together by setting the direction for all the body work. It is the basic position from which all the standing exercises start and will be referred to later as the basic orienting exercise.

Exercise 13 / Basic orienting position

If you are in a group, the group should form a circle. Each person stands facing inward with his feet about 8″ apart and parallel.

Lean forward so that the weight of your body is on the balls of your feet. The knees are slightly bent, the pelvis hangs loosely and tilted backward, the upper part of the body is straight and relaxed.

Let your belly out and take four or five deep and audible belly breaths. Drop the pelvic floor (as if you were going to defecate or urinate). If you have qualms, go to the bathroom first. You will find that the anxiety of "letting go" is there whether you have to "go" or not.

Now breathe easily and deeply and try to sense how much you can "let down" into your feet.

• Was your breathing free, easy, and deep? Could you feel the respiratory movements in the abdominal cavity? Could you let out a sound?

• Were your knees bent and your feet parallel throughout the exercise?

• Was your pelvis back and loose? Could you feel any tightness in your ass or tension in the pelvic floor?

• Could you feel the weight of the body settling down into the balls of the feet?

If all the weight of the body rests on the balls of the feet, there will be no needless holding oneself up in other parts of the body. The shoulders will drop and the chest will become soft. The pelvis will hang loosely and tilted backward in the proper position. Unfortunately, this is not easy to do. We are unconsciously afraid to "let down," and so we hold ourselves up by tensions in many parts of the body. These tensions prevent us from letting down. We hold our shoulders up because we don't feel our feet on the ground, and we hold our jaws tight because we are afraid to cry. We are afraid to "let go" in the anus for fear of soiling ourselves, and so we hold our asses tight. However, with awareness and practice we can arrive at a point where we are able to feel our legs and feet holding us up and the rest of the body at ease and flowing.

Exercise 14 / The "joy of being alive" stretch

From the position described above, stretch the arms forward, up-

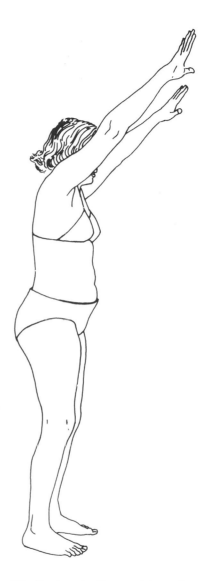

Fig. 19. Basic orienting position—the "joy of being alive" stretch

61

ward, sideways, and down, keeping the palms turned away from the body. Do this slowly, coordinating the breathing with the arm movements. When the arms stretch forward, breathe out with an audible sigh. Breathe out again when the arms are in the upright position, again in the sideways position, and once more in the downward position.

Repeat the exercise several times.

The orientation in bioenergetics is to have your feet firmly planted on the ground. At the same time you should be able to reach toward the sky. Thus, the direction is both downward and upward. The exercises involve two actions, grounding or "letting down" and stretching or "reaching out." When you reach out or up, your feet should not lose contact with the ground. Your legs should not become tense, and your pelvis should not be pulled in or up. Reaching is stretching, not tensing.

Warming-Up Exercises

A warming-up exercise should generally precede the more directed and intense body work. Any one or two of the following will serve.

Exercise 15 / Shaking loose

Stand with feet parallel and about 8″ apart, knees slightly bent, weight forward, pelvis loose, back straight. The arms should hang loosely at the sides.

Rapidly bend and straighten the knees so that a bouncing motion is produced without the feet going off the ground.

This action should produce a shaking in the whole body that affects the breathing so that it sounds like a dog panting.

Continue this exercise for about a minute, then rest with knees still bent and breathe easily.

• Did your whole body shake with the exercise?
• Was your breathing harmonized with the motion?
• Did you allow your upper body to lean backward, pulling you out of

your feet even though your knees were bent?

Exercise 16 / Slow jumping

From the same position, jump slowly on both legs, just barely lifting the toes from the floor. Each jump should take about a full second.

Continue this exercise until the legs feel tired. Then rest in the standing position with knees bent, weight forward, back straight.

• Are you out of breath? Are your legs vibrating? If so, that's good.

This is a strong exercise, and most people who do it correctly find themselves out of breath. Let yourself breathe easily.

Exercise 16-A / Variation

Jump easily twice on each leg and then change to the other. The rhythm is 1, 2, change legs, 1, 2, change legs. It is much less strenuous than the above.

Exercise 16-B / Variation

Skip or stamp around the room, preferably in a circle, swinging your arms.

Exercise 17 / Jumping rope

In Mrs. Lowen's classes jump ropes are available. When people come in they do some jumping to loosen and warm up. Not every floor can take a large group jumping, however, so the jumping should be staggered, i.e., one or two persons taking turns jumping.

Exercise 18 / Rocking back and forth on the feet

In the standing position, rock backward and forward on your feet. Lift the heels slightly when rocking forward, then lift the toes when rocking backward. Breathe easily and deeply. Keep the knees bent, the pelvis hanging loosely.

Standing Exercises

The purpose of these exercises is to mobilize sensation in the legs and feet so you feel your feet on the floor. These exercises will also induce a strong vibration in the legs, which will gradually extend upward to include the pelvis and upper part of the body. The vibration loosens tensions and helps you sense your legs.

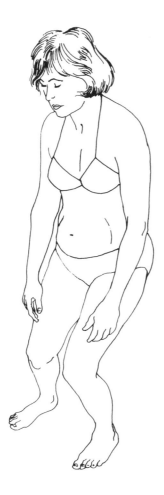

Fig. 20. Standing with weight on one leg and toes of other touching the ground

Exercise 19 / Weight on one leg with knee bent

Assume the position of the first orienting exercise: feet about 8″ apart and parallel, weight forward, pelvis tilted back and loose, belly out, body straight and relaxed. Drop the pelvic floor. Bend the left knee and shift all your weight onto the left foot. The right foot should rest flat on the floor. Breathe easily and deeply. Hold the position until you get uncomfortable.

Then shift your weight to the right leg, bending the right knee.

Repeat again on each leg. Then return to the resting position.

• Can you let down into each foot or do you feel a stiffness in the knee?
• Did you hold your breath?
• Are your legs vibrating strongly? Once they begin to vibrate, much of the pain and tension is released.
• Are you frightened that your knee will buckle under you and that you will fall?

This fear, called falling anxiety, will cause you to stiffen and tense the knee, thus placing an excess strain on its ligaments. Falling anxiety can be worked on through the falling exercises described below.

Exercise 19-A / Variation

Here is a stronger version: shift all your weight onto the left foot and bend the left knee more fully. Now lift the right foot off the ground, but let the toes touch the ground for balance. This position places more stress on the left leg and allows you to feel it more intensely. Hold the position until it becomes painful.

Make sure that your pelvis is tilted back and loose, your pelvic floor dropped.

Breathe easily and deeply.

Repeat the exercise on the right leg.

Exercise 19-B / Variation

If you are more advanced, a much stronger variation of this exercise can

Fig. 21. Weight on one leg and other off the ground

also be used. One leg is placed forward, the knee fully bent and all the weight allowed to rest on that leg.

The other leg is lifted off the ground. The body is bent forward so that the hands are close to the ground to provide a balance if needed. The pelvis is back.

This exercise should be done with a thick rug or folded blanket under the knee so that you can let yourself fall when the strain becomes too great.

Repeat the exercise on the other leg, again letting yourself fall when the pain becomes strong.

Repeat once more on each leg.

Return to the orienting position.

• These stronger exercises create more sensations in the legs. When you return to the orienting position (weight forward on the balls of the feet, knees slightly bent, pelvis back, pelvic floor relaxed), you should find that your legs are vibrating intensely. Is this happening?

Repeat Exercises 4 and 1

At this point, we would like to reintroduce exercise 4, the bow or arch, described in chapter 2. Please turn to page 20.

This exercise should now be followed by exercise 1, the basic vibratory and grounding exercise, described in chapter 1. Please turn to page 11.

Exercise 20 / Deep knee bend—crouch

The previous exercise can be followed by this one, which will bring you closer to the ground.

Bend both knees fully so that you are in a crouch. Lift the heels slightly so that the weight of your body is on

Fig. 22. Crouch

the balls of the feet. Extend your arms forward and lean forward as if you were about to dive into a pool for a race. Hold this position for one minute or so, breathing deeply into your pelvis.

• Can you feel the breathing movements in your pelvis?
• Can you feel the tightness in your lower back?
• Are you holding your shoulders rigidly?

If your knees and ankles are stiff, you may not be able to bend low enough to get into the crouch. This means you need more work with your legs to loosen them.

Exercise 21 / Moslem prayer position for rest and deep breathing

When you are tired, let yourself go down on your knees. Stretch forward, your hands in front, palms

Fig. 23. Knee-chest position

against the floor, and rest your forehead on your hands. Your elbows should be spread apart.

Arch your back so that the belly is let out as fully as possible.

Breathe easily and deeply into your lower belly.

Let this be a position of rest for about a minute or two. It is similar to the position used in Moslem prayers.

In this position you may feel the anus opening and closing with each breath. Have no fear that you will mess; your internal sphincter remains closed.

If you need a further period of rest, lie flat on the floor on the belly, just letting the body be and sensing the different parts that make contact with the floor. This is as close to the ground as one can get while still alive.

Exercise 22 / Working with your ankles

In most people tension in the ankles cuts off feeling from the feet. It is necessary therefore to loosen the ankle so that it can be easily flexed. This is an important bioenergetic exercise. It starts from a position on the knees. Thus, it can be done immediately following exercise 5 or any

Fig. 24. Ankle position

other time. The toes are out-stretched.

Place the left foot flat on the ground about 4'' to 6'' behind the right knee.

Shift all the weight to the left foot, letting the right knee move sideways if necessary.

Keep the weight of the body on the ball of the foot stretching both arms forward for balance. Rock forward and back on the left foot, allowing the left heel to come off the ground only as necessary. Press down on the left foot as you rock forward.

Keep the left knee in line with the big toe of the left foot.

Repeat this exercise with the right foot, placing it about 4'' to 6'' behind the left knee. Shift the weight to the right foot and proceed as above.

The exercise should be repeated on each foot.

• Can you feel the tension in your ankle?

• Do you feel a stretch of the Achilles tendon (the tendon joining the muscles of the calf to the bone of the heel)?

• Do you feel your foot pressing into the ground?

• Could you relax your calf, thigh, and upper body as you pressed into your foot?

Exercise 23 / Extending the foot

This is another ankle exercise. It is done from the same position as the preceding one. You get onto both knees with your legs and feet

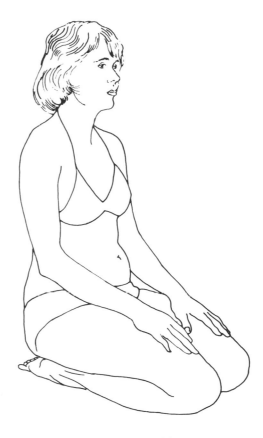

Fig. 25. Sitting on ankles

extended behind you.

Sit back on your extended feet. Some people can do this very easily; others have considerable difficulty.

Exercise 24 / Extending the thigh

This exercise is done from the position of exercise 23. You are sitting on the heels of your extended feet.

Place your fists against the soles of your feet. Press down with your fists; this should relieve some of the tension in your ankles and feet.

With your weight on your fists, raise your hips, extending the thigh. Drop the head backward, letting the body arch. This will stretch the muscles of the thigh, which are quite contracted in most people.

It is important to let the pelvis hang loosely.

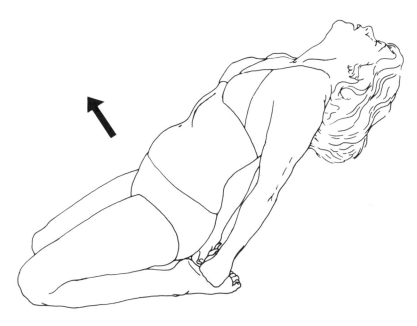

Fig. 26. Lifting off ankles

Hold the position as long as you can, then relax and try it again.

• Does this exercise cause you pain in your feet? Such pain indicates the presence of tension in the foot muscles.

• Can you sense the contraction in the thigh muscle? Is it painful?

• Are the arms painful?

Exercise 25 / Flexing the foot— return to crouch

The extension of the foot can be reversed by flexing the foot. Following exercises 23 and 24, you should come up on the feet into a crouching position.

Extend your arms as far forward as you can, touching the ground lightly.

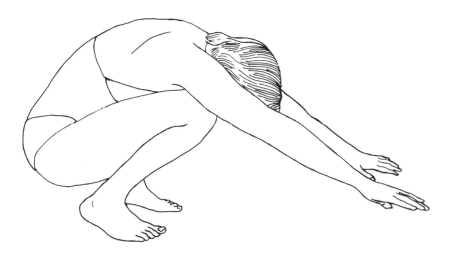

Fig. 27. Deep knee crouch (flexing foot)

Lift your heels very slightly from the floor, letting your weight move onto the balls of your feet. Hold. Rest by letting yourself go onto your knees.

• In the crouched position can you feel yourself deep inside the pelvis? It is good to hold this position for about half a minute to sense the depth of your breathing.

• Does the crouched position help you feel your ass? Remember that this is the position people used for defecation before the flush toilet become popular.

• Can you relax your back?

• Can you feel your belly?

Exercise 26 / A stress exercise for the legs

This is designed to make you fall and thereby give up the rigidity in your knees. The exercise starts with you resting on both knees.

Place the left foot flat on the floor about 6'' in front of and parallel to the right knee.

Reach forward and place the fingertips of both hands on the floor about 18'' in front of the foot.

Shift your weight forward onto your fingers. As you do this, *lift* the right foot off the floor. If this is too much weight for the fingers, you can use your palms for support. Bend the left knee so that it is about 8'' above the ground and bring the right leg (which remains off the floor) up close to the left one.

Hold this position as long as you can without forcing, then let your left knee collapse onto the floor. You should be working on a rug or a folded blanket you can fall on.

Go back to the original position with both knees on the floor and repeat with the right leg forward, following the same directions.

• Normally a person should be able to hold the position about thirty seconds, but not much longer. Did you collapse immediately? *Note:* The immediate collapse of the knee denotes that the leg cannot hold the person up unless the knee is rigidly locked. Overweight people have great difficulty with this exercise, but only partly because of their excess weight. Without going deeply into the complex psychology of over-weight people, we believe that part of their problem is connected to a feeling of insecurity about the ability of the legs to support them. Their basic oral insecurity is compensated by overeating. This exercise helps to focus on this problem. Also, doing it regularly will help strengthen a person's legs by bringing more feeling into them.

• Did you hold the position for a minute or more? If you did, it may mean that you need to prove how well you can take stress. In that case, you may leave yourself open to a greater stress than the body can tolerate. This exercise is not designed to test your performance under stress. None of the exercises in this book are. Here, the important thing is how much feeling you get of your legs and feet.

• Were you breathing during the exercise or did you hold your breath? Holding the breath in this exercise

Fig. 28. Falling exercise—deep knee bend

denotes a fear of falling that has no relation to the situation, for there is no danger if you fall here.

• Could you feel your left foot actually pressing against the floor to hold you up?

• Do you sense that the moment you feel all the stress and pain you tend to straighten the leg to avoid falling?

Repeat the exercise with your right leg. You may notice that one of your legs is stronger than the other. This is a very common situation.

We recommend that the exercise be done twice on each leg. Most

Fig. 29. Falling exercise—straighten leg

people find it to be somewhat easier the second time. They are less frightened of the stress, knowing that they can fall safely.

Repeat Exercise 13 / Basic orienting position

Generally, after the previous exercise, we ask people to stand in the orienting position, that is, knees flexed, weight forward, belly out, back straight, pelvis loose, and shoulders and bust down. They are then advised to breathe easily and deeply into the belly and to focus their awareness on their legs and feet. Please turn to p. 60.

- Are your legs vibrating?
- Can you sense the flow of excitation into your legs and feet?
- Are you more aware of your feet? Do they feel warmer to you? The warmth results from an increased flow of blood.
- Can you feel your feet to be more in contact with the floor?

Exercise 26-A / Variation

This variation is very helpful in giving a person the sense that his legs can lift him up and hold him up. The position and the procedure are similar to exercise 26. You should be forward on your left foot with the knee bent as above and the weight largely on your hands.

Push your body back onto the heel of your left foot, which now rests on the floor. Although your hands are still touching the floor, there should be no weight on them. Press down on the left foot partly straightening the left knee until you feel the left leg supporting you and carrying your weight.

Repeat the exercise by letting the left knee bend again, shifting your weight forward onto your hands until you are in the original position. Then push back again as above until all your weight is transferred to the left foot.

Repeat the exercise on the right leg just as above.

- Can you now sense the ability of your legs to hold you up?
- Do your legs feel more "there" for you?

Exercise 26-B / Variation

In this version, the person rises completely on each leg in the same manner as above, taking his hands off the floor. As soon as he is upright, both feet should be on the floor.

Exercise 27 / Shaking a leg loose

From the relaxed standing position described in exercise 13 (see p. 60), extend one leg off the ground and flip the ankle repeatedly to shake loose any tension. Do the same with the other leg. Keep the standing leg bent at all times.

Exercise 28 / Kicking out with a leg

As in the above exercise, start from the relaxed standing position and extend one leg forward off the ground. Give a number of kicks, pushing with the heel of the extended leg, keeping the knee of the supporting leg bent. Then repeat by changing legs.

Now we will list a few exercises that need little explanation in order to show the range of maneuvers for the legs and feet to make them come more alive. Before we proceed, it should be emphasized that the exercises are not to be rushed through nor performed at the maximum stress possible. Also, there should be enough time between each exercise for a person to sense what is going on in his body. In bioenergetic philosophy *doing* is less important than *feeling*. Many people do exercises mechanically, as if in some calisthenic class, so as *not* to feel. This defeats our goal of aliveness.

Exercise 29 / Exercise for the soles of the feet

Almost every person in our culture can be described as a "tenderfoot." Unlike those of primitive people who walk barefoot, the soles of our feet are too sensitive. In part this is due to their being protected by shoes. In larger part, however, it results from spasticities in the muscles of the soles. Our arches are either flat or too high. To release these tensions we recommend standing on a wooden dowel 1'' in diameter. In the absence of a dowel the handle of a tennis racket will do.

Place the dowel on the floor, and put one foot on it. Let the weight of your body rest on that foot as much as you can. Then move the foot on the dowel, exposing different parts of the sole to the pressure. Make sure you are breathing during the exercise.

Repeat the exercise with the other foot, placing the dowel at different parts of the sole.

Most people experience pain doing this exercise. The pain is due to pressure on a tense muscle. If you stay with the pain, the muscle relaxes and the pain diminishes. With continued practice you will find the pain greatly reduced. It is not masochistic to submit to pain in the interest of better health, a more alive body, or a more relaxed foot. That is reality. Masochism exists when the submission to pain leads nowhere. *

Exercise 29-A / Variation

A simple lifting exercise is also useful. Stand in the relaxed position, weight forward on the balls of the feet and knees bent. Keeping the knees bent, slowly lift the heels off the floor, then let them come down to touch the floor. Keep repeating this procedure.

• Can you feel the muscles of the soles of the feet working to lift you up?

This exercise is expecially useful for flat feet.

Exercise 30 / Jumping in the crouch position

From the crouched position described in exercise 20, p. 66, jump easily up and down on the balls of the feet.

This exercise will enable you to sense the active pressure of your feet against the ground.

Exercise 31 / The mule kick

This is a good exercise to loosen the hip joint and stretch the muscles in the back of the legs. Assume a position on elbows and knees. Lift the left leg off the ground and bend the left knee so that the leg is close to the body. Thrust the heel backward using as much hip power as you can. Make the kick as straight and parallel to the floor as you can. Do a number of kicks with the left leg, then repeat the exercise with the right leg.

• Did you follow through with the kicks or did you hold back in the knee at the end?

*A fuller discussion of the relation between masochism and pain can be found in Alexander Lowen, *Pleasure* (New York: Penguin Books, 1975).

• Could you feel the kick coming from the hip and did you sense the power in the hip?

• Could you relax the upper half of your body?

Exercise 32 / Stretching the hamstring muscles

By now, the hamstring muscles at the back of your thighs and knees may be very tight. If so, here is an exercise to stretch them.

With feet parallel and about 12″ apart, lean forward, bending your knees until your fingertips touch the floor. This position is identical with the one in exercise 4, p. 20.

Move hands forward until the heels are off the ground.

Now straighten the knees so that the heels press against the floor.

Hold the position for a few mo-

Fig. 30. Mule kick

ments, then release the knee and start the stretch again.

Exercise 33 / The bear walk

The bear walk is an excellent exercise to stretch the hamstring muscles, which are very tight in many people. Come down on all fours with your hands and feet flat on the floor. Walk around the room keeping hands and feet flat on the floor.

If you understand the principles of bioenergetics, you can improvise many exercises in the standing position to help you get more in touch with your body and relieve its tensions. The important thing is to know that you are working toward achieving greater feeling.

Working with Hips and Pelvis

Grounding a person so that he feels his feet solidly on the floor is a prerequisite for releasing the tensions in the hips and pelvis. The pelvis will not move naturally, that is, freely and spontaneously, unless it is suspended between the head and the feet. This is the principle of the bow. With a violin or guitar string both ends must be securely anchored for the string to vibrate rhythmically and produce a sound. The same thing is true of a bow. When the bowstring is drawn, the bow becomes charged and all that is necessary to project the arrow is to release one's hold. The arrow will fly without pushing by virtue of the charge imparted to the bow.

This principle applies to the movement of the pelvis. If the feet are fully in contact with the floor, it is only necessary to pull the pelvis back to create the charge that will spontaneously move it forward. The energy for the charge is produced by the metabolic processes of the body in connection with the breathing. Thus, any tension in the body that restricts the breathing or prevents grounding limits pelvic motility.

When the pelvis is pulled back, it is in the charge position, provided that the whole body is properly balanced on the front of the feet, as in our first orienting exercise. In the forward position, the pelvis is already discharged. It cannot move forward spontaneously but has to be pushed. Pushing tightens muscles and so reduces the flow of excitation and pleasure. It is to be avoided in these exercises as well as in sexual intercourse.

Pelvic tensions develop to limit sexual feeling. It will be difficult, if not impossible, to free the pelvis if one is inhibited about sexual feeling. These exercises do not necessarily create sexual feeling, but they can make one aware of this feeling if it is present. However, sexual feeling is not the same as genital excitation, which is a focus of sexual feelings upon the genital apparatus. That does not happen with these exercises.

The following exercises are designed to help you sense any tensions in the pelvic area. Some release of tension will occur through doing them. In most people these tensions are quite severe, and we use many other exercises to loosen them up. Any exercise that mobilizes the lower part of the body, whether from a standing or lying-down position, will affect the pelvis. Kicking is one such exercise. In addition, there are the specifically sexual exercises that will be described later.

Exercise 34 / Side-to-side pelvic movement

Stand on both feet, which should be parallel and about 12'' apart. Bend the knees about halfway.

Shift the weight to the left foot without moving the right one and swing the pelvis to the left side by pressing on the left foot.

The upper part of the body stays relatively straight and is inactive.

Slowly shift the weight to the right foot and swing the pelvis to the right side, pressing down on the right foot.

Repeat the exercise to the left and right a number of times slowly, letting the pelvis swing as the pressure goes into each foot.

• Are you letting yourself breathe into the belly?

• Do you sense the movement of the pelvis coming from the ground as you press down on each foot?

• Do you feel any tension in your lower back?

• Were you able to keep your body weight forward as you shifted from side to side?

Exercise 35 / Circular pelvic movement

This exercise is very similar to the hip movements in a hula dance. The position is the same as in exercise 34.

Shift your weight to the left foot, and pressing down on that foot, let the pelvis move to the left.

Let the weight of your body come back to the balls of both feet, and, as it does, swing the pelvis forward.

Shift to the right foot, and by pressing on that foot, swing the pelvis to the right.

Return your weight to both feet, and keeping forward on the balls of your feet, pull the pelvis back.

Do this exercise as a continuous movement several times. Then repeat it by reversing the direction so that the pelvis moves in a counter-clockwise direction.

• Are you breathing with the belly? If you push the pelvis forward, you will contract the abdominal muscles and block any belly breathing.

• Can you feel the weight of your body shift on your feet as the pelvis swings in a circle?

• Can you keep your knees bent and loose or did you straighten them? It is important to stay "low down."

• Was the upper half of your body fairly straight and inactive during the movement? This indicates that you were able to direct your energies properly.

Exercise 36 / Forward and backward pelvic movement

This exercise is done from the same position as the first one. Keep the weight of your body on the balls of the feet.

Pull the pelvis backward by arching the lower back, but remember to keep your weight forward. This will prevent an exaggeration of the back arch into a lordosis (an exaggerated concavity of the lumbar spines).

Let the pelvis swing forward by pressing on the balls of the feet and breathing out.

Exercise 37 / Cocking the pelvis

This exercise is designed to give you a sense of the power available for thrusting when the pelvis is pulled

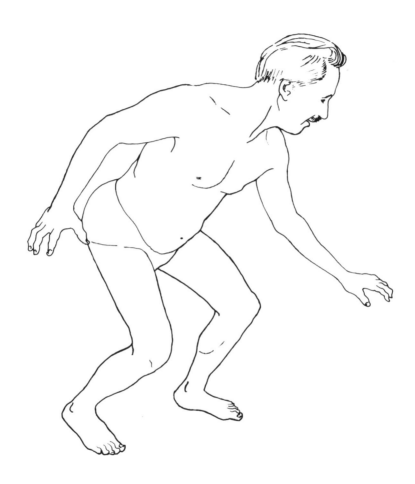

Fig. 31. Cocking the pelvis

back. This exercise is also done from the same position as the previous ones.

Put your right foot forward with your weight on it, and bend both knees. This will cause your left heel to come off the ground.

Bend your trunk downward, extending the right arm and pulling the left arm backward.

Cock your ass backward and upward, and shift your weight to the ball of your right foot. You are now leaning forward in an aggressive position.

Try the same exercise with your left leg forward. Most people find it easier to do with the left foot advanced because we tend to start walking by advancing the left foot.

• Did you feel any tension in your right knee when that foot was forward? In the small of your back?

• Could you sense how the feeling of aggression was increased when you cocked your ass? Aggression has the positive meaning in bioenergetics of "moving toward."

Pull the pelvis backward with an inspiration of air, then, again, let it swing forward as you breathe out.

Keep a backward and forward movement going easily as you breathe in and out.

• Were you able to coordinate your breathing with the pelvic swings?

• Did you have to force the pelvis forward by pushing from your ass or pulling from the front? In the former case, you had to tense your buttocks; in the latter, you contracted the abdominal muscles.

• Were you able to keep the weight of your body at all times on the balls of your feet?

Exercise 38 / Duck wiggle

This is a more difficult exercise since it depends upon loose hips and pelvis. Start from the same position as in the previous exercise, but bend both knees fully, keeping feet flat on the floor.

Shift weight to the front of the feet without lifting the heels.

Bend forward and cock the ass backward.

Move the ass from side to side without swaying on legs or moving the upper trunk. Weight should remain evenly balanced on both feet.

• Could you move the ass freely by itself?

• Did you keep your weight forward?

• Were you able to breathe easily during the exercise?

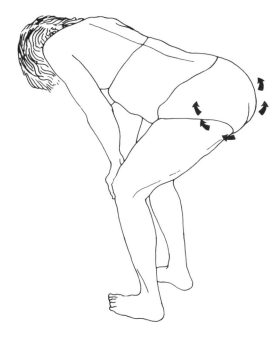

Fig. 32. Duck wiggle

Working with Arms and Shoulders

To free up the arms and shoulders often requires the use of expressive exercises such as hitting, twisting the towel, and the like. These are described in a later chapter. They can be included in the class work if appropriate. The exercises that follow involve more generalized movements.

Exercise 39 / Swinging each arm

Start the exercise from the orienting position, exercise 13, on page 60.

The body weight is forward, the knees are slightly bent, and the belly is out.

Fully stretch the left arm backward

and up, bending slightly forward.

As the left arm goes backward, the right arm extends straight downward and a little forward.

Make a *slow* circle with the left arm upward, forward, and down, keeping the arm as fully stretched as possible to involve the shoulder.

Do the same with the right arm, extending it fully backward while the left arm remains extended downward.

Repeat left and right for a number of times.

This exercise can also be done by starting from a forward stretch of the arms and circling backward.

• Could you feel the whole shoulder move while you did the exercise?

• Did you feel any tension around the shoulder joints? Along the sides of the body?

• Were you able to breathe as you did the movement?

Exercise 40 / Two-arm swing

From the same position as in the preceding exercise, extend both arms sideways.

Swing both arms forward and downward to the sides.

Breathe *out* audibly during the forward and downward swing.

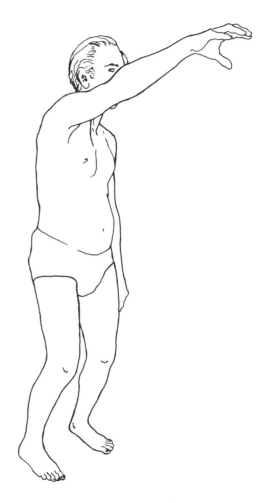

Fig. 33. Arm stretch

85

Breathe *in* as you bring the arms up sideways for the next swing.

Keep doing the swings, moving and breathing faster with each swing.

Exercise 41 / Flying like a bird

This exercise is similar to the preceding one, but the movement stays in the plane of the body, with the arms remaining extended sideways.

Extend arms sideways and flap them up and down.

Lean forward and move the arms faster and faster until it feels as if you would take off like a bird.

Let arms fall to the sides and rest.

Exercise 42 / Shoulder roll

Again, from the same position, with arms hanging loosely at your sides, bring the shoulders up.

Now move the shoulders forward, downward, and backward, making a number of circles.

Repeat in the opposite direction.

Shoulder exercises can also be done in the sitting position and will also be listed in that section for convenience.

Exercise 43 / "Get off my back"

This is an expressive exercise routinely used in class work. Most people find considerable satisfaction in doing it. Start from the same position as in the preceding exercises.

Bend the elbows and then raise them up to shoulder height. This will serve to extend the upper arms.

Forcibly thrust both elbows backward and say "Get off my back."

Repeat the exercise a number of times, giving vocal expression to a feeling of anger.

• Could you feel that this exercise straightened up your back?

• Were you aware of being somewhat bowed as if you were carrying someone on your back? Most people greatly enjoy doing this exercise, which means that they feel someone on their back.

Exercise 44 / Punching forward

This is similar to the preceding exercise. Bend your elbows and raise them to the level of your shoulders.

Make two fists with the thumbs on the *outside.*

Drive both fists strongly forward and say "Get away."

Repeat a number of times.

Exercise 45 / Driving your fists down

Bring both fists as close to your armpits as possible.

Drive both fists downward alongside the body with a grunt.

Raise, and repeat the exercise several times.

• Did you feel a wave pass through your body as your fists came down?

• Did you keep your knees bent and did the movement seem to drive your feet into the ground?

• Did you feel some vibration in your head?

Exercise 46 / Shaking your fists

Assume the same position as above.

Raise both fists until they are in front of your face.

Shake them forcibly and say "no" vehemently a number of times.

Let your voice be loud and clear.

• Could you sense if you had an angry or frightened expression on your face while doing this exercise?

• Was your voice strong and assertive?

• Were you able to keep a forward stance or did you tend to lean backward?

Working with Head and Neck

These exercises are designed to loosen some of the neck tensions and free up the head. If your head is held very rigidly on the neck, you may experience some dizziness from the exercise. If this happens, stop and wait. Closing your eyes during the exercise may prevent the dizziness by eliminating the moving field of vision.

Exercise 47 / Stretching the neck

Place your hands on the back of the *head*, intertwining your fingers.

Press downward with the hands letting the head yield fully under the pressure.

Keep your knees slightly bent and your back straight but not rigid. Your weight should be forward. Breathe deeply. See fig. 34 on the next page.

• Could you feel the stretch in your neck? Did you feel any pain

Fig. 34. Neck exercise (standing)

down your back or in your shoulders?

• Did you sense yourself standing straighter when you stopped?

Exercise 48 / Neck massage

With your hands in the same position as in the preceding exercise, use your thumbs to feel and massage the muscles that join the head to the neck. Your head should be bent forward.

• Could you sense how tight these muscles were?

Exercise 49 / Forward snap

In this exercise the head is lifted and pulled back, then allowed to fall forward with a grunt. Do this exercise very easily at first until you feel fully at ease with it.

Exercise 50 / Rolling the head

Let your head fall forward lightly, then roll it in a circle from left to right, breathing slowly and easily as you do so. Keep your eyes open, focusing on different objects that pass your line of vision. Blink your eyes often. Do this at least three times, then repeat in the opposite direction.

Let your shoulders hang as low as possible.

If you become dizzy, stop the movement and bend over as in the grounding exercise (page 11), letting your feet and fingertips make contact with the ground. This exercise can also be done in the sitting position, which provides more security.

• Were you breathing easily?

• Did you hear any cracking sounds in the neck? Do not worry; these are caused by the release of pressure between the articular surfaces (surfaces between the joints) of the neck vertebrae.

Sitting-Down Exercises

The sitting-down position is very useful for exercises involving the upper half of the body. However, these exercises will only be valuable if you sit correctly: the back straight, the legs crossed, the belly let out, and you feel grounded through the ischial tuberosities of the buttocks as shown in figure 35 (see p. 90). In this position you should feel your bottom pressing firmly against the floor with the weight of the body carried forward. Sitting back on the tail bone, as so many people do, results in collapse of the back by forcing the pelvis forward and tightening the abdomen. This prevents abdominal respiration and blocks the flow of feeling the exercises are designed to produce.

The sitting position provides more security than the standing one, and so it allows one to "let go" of the head more easily. It is therefore the position of choice for exercises involving the neck muscles. It is also an excellent position for the eye exercises, which will also be described in this section.

Exercise 51 / Grounding yourself in the sitting position

This is the basic position from which all the other sitting-down exercises are done.

Sit cross-legged on a rug or on a folded blanket. Lean forward, arching the back until you feel the ass pressing against the floor. Let the belly out. Let your arms rest comfortably upon the knees or thighs.

Hold the head up loosely. Breathe easily and deeply for a minute or two, and try to sense the breathing movements in the belly and ass.

• Were you able to assume the correct position? Is your chest soft and moving gently with your breathing?
• Can you feel the respiratory movements in the belly?
• Are you aware if your anus is pulled in tight or relaxed? Can you let it out? Do you feel any anxiety doing this?
• Do you sense any tension in your back or in your legs?
• How long can you remain comfortably in this position?

Mabel E. Todd says of this position: "If the body-weight is balanced upon the tuberosities of the ischia [see figure 35] and there is no pull forward by the leg muscles, the planes of force acting upon the key-stone of the pelvic arch will be identical to those in the balanced standing position." * Since the body in standing is balanced upon the balls of the feet, we refer to the ischial tuberosities as the balls of the ass.

*Mabel Elsworth Todd, *The Thinking Body* (New York: Paul B. Hober, Inc., 1937), p. 281.

ischial tuberosity

Fig. 35. Sitting position

Being properly balanced is the first step to being grounded. The next is to breathe freely and deeply.

It is also interesting to note that this position is similar to the lotus position used in Yoga and Zen meditation.

Exercise 52 / Relaxing the muscles of the waist

The waist, like the neck, serves to connect two major segments of the body in such a way that each segment can turn somewhat independently of the other. Thus, the head can turn left or right because of the flexibility of the neck. If the neck becomes too rigid, this movement is hindered. Similarly, the torso can twist to each side because the waist is flexible. Rigidity not only restricts the movement, it also interferes with the flow through the connecting segments, causing a decrease in the feelings of being unified or integrated.

From the cross-legged sitting position, place the right hand on the left knee.

Twist the body to the left so that you can look over the left shoulder.

Hold the position, taking a number of deep breaths, and then slowly face forward.

Place the left hand on the right knee, twist to the right, and look over

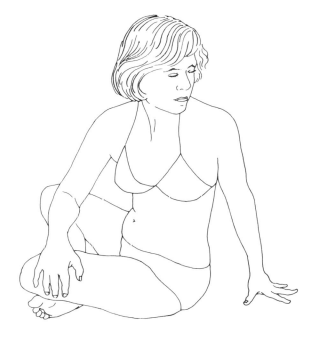

Fig. 36. Waist twist

the right shoulder. Hold the position for a few moments, breathing deeply. Slowly return to the starting position.

- Could you feel any tension in your shoulders, waist, back, or hips?
- Was it difficult to breathe down into your belly?

Exercise 53 / Arm stretch

Sit with your back straight and stretch both arms out to the side, keeping shoulders down.

Raise both hands with palms facing outward.

Push outward with the heels of the hands to stretch the arm muscles.

Don't lock your elbows.

Exercise 54 / Charging the hands

This exercise is one of the most dramatic in the bioenergetic repertoire since it generally produces some unusual and vivid sensations in the hands. It is done from the same position as the previous exercise.

Place the tips of the fingers together. Spread the fingers apart and press them against each other, keeping the palms apart.

Turn the hands inward so that they point toward the chest.

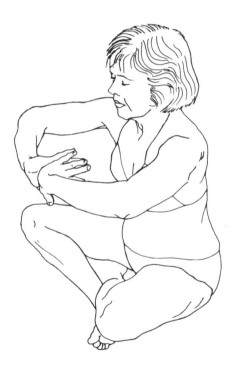

Fig. 37. Hand stretch (charging hands)

Holding the fingers against each other and keeping the palms apart, move the hands outward as far as possible.

Breathe easily and deeply for a full minute before relaxing the hands.

Relax by holding hands loosely in front of the body, fingers lightly extended, and look at your fingertips for thirty seconds. Drop your shoulders and keep breathing.

Next, cup the hands loosely and slowly move them toward each other until they are 2″ apart.

• Did you have the sensation of something between your hands as you brought them together?

• When your hands were forward and relaxed, did you feel any tingling in your fingers?

• Did you feel any tightness in the chest muscles while doing the exercise?

This exercise makes it possible to experience the energetic charge in the hands. When this charge is strong, it radiates from the hand and fingers creating an energetic field around the hand. As the hands approach each other in the charged state, people generally sense some-thing substantial between them. I believe that this "substantial some-thing" is the energy field between the hands. It can be demonstrated by Kirlian photography.*

Exercise 55 / Shaking hands loose

Hold your arms outstretched with hands loose.

Shake your hands vigorously to loosen them.

Rotate your hands in circles slowly, first inward toward each other for a number of times, then outward away from each other.

Exercise 56 / Finger stretch

Place your hands flat on the floor alongside your body with fingers spread apart.

Lean forward and stretch each hand between the little finger and the thumb.

Repeat the stretch, placing pressure on the fourth finger and the thumb.

Do the same with the middle finger.

Repeat with the index finger and thumb.

---

*Kirlian photography is a method invented by the Russians. If a hand is placed in a strong electro-static field and photographed, the plate will show what looks like a radiation from the hand.

Exercise 57 / Wrist exercise

With your hands flat on the floor facing forward, lean forward until the heels of the hands are slightly raised.

Rock back and forth, pressing down on the hands.

Place the back of the hands on the floor, and rock backward and forward to stretch your wrists.

Exercise 58 / Shoulder-loosening movements

From the relaxed sitting position, several exercises can be done to relax the shoulders.

a. The shrug. With arms hanging loosely at your sides, slowly lift up the shoulders, then let them drop as if to shrug off the cares of the day. Repeat several times, breathing in as you raise the shoulders and exhaling as you let them drop.

b. The shoulder roll. With arms extended sideways, slowly roll the shoulders up to the ears, forward, down, and then back several times. Inhale as you draw the shoulders back and up; exhale as you bring them forward and down. Now reverse the direction, bringing the shoulders back and up first, then forward and down. Maintain the same breathing pattern. The downward and forward movement should coincide with a collapse of the chest, not the back.

c. Reaching out. Most of us reach with arms and hands, not with our shoulders. This exercise will help you realize the difference. First, reach out with arms and hands. Now, slowly extend your reach by bringing your shoulders forward as you let the air out. When the expiration is completed, slowly draw your shoulders back as you take a breath; then slowly reach forward again with hands, arms, and shoulders while breathing out. Repeat this several times.

• Can you feel your chest soften as you reach out, and can you sense that the movement of reaching seems to come from your heart?

d. Extending the shoulders. Stretch your arms out to the sides. Raise your fingers and push with the heels of your hands to increase the stretch of your arms and shoulders. Hold the stretch through a number of breaths, then let your arms drop slowly to your sides. Try to keep your shoulders down while doing this exercise. Repeat it two or three times.

Exercise 59 / Neck-loosening exercises

Now is a good time to loosen the neck muscles. The exercises in the preceding section, "Working with head and neck," can be used and/or the following exercises, which were not given before.

Bend the head forward and back.

Circle the head clockwise and counterclockwise.

Tilt the head to either side. The head is tilted first to the right, and the face is turned upward; then the head is tilted to the left and the face is again turned upward. This may be repeated several times.

Exercise 60 / Stretching the neck muscles

This exercise is similar to one in the preceding section, "Working with head and neck," but is here done in the sitting position.

Place the hands on the back of the head and interlock the fingers.

Pull the head downward, gradually increasing the pressure.

Bend forward and continue the pressure to feel the stretch extending down the back.

Return to original position but keep hands on the head.

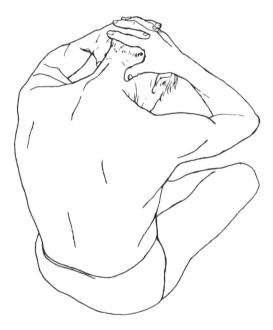

Fig. 38. Neck exercise (sitting)

Place the thumbs on the muscles at the back of the neck and slowly massage them, working from the root of the neck to the base of the skull. Also massage along the base of the skull.

• Could you feel the tightness of your neck muscles? In most people they are like tight bands alongside of the spinous processes of the vertebrae.
• Could you feel the tightness of the muscles at the base of your skull? This tightness is a common cause of headaches.

Exercise 61 / Eye exercises

The eye muscles, just like those in other parts of the body, are subject to tension. Relaxing these muscles helps keep the eyes soft. Two different exercises are used to accomplish this, one involving eye contact with the other members of a class, the other involving eye movements. It is important not to let your gaze become fixed or rigid. Do not stare. Staring freezes the eyes and prevents contact and feeling. Allow yourself to blink during the exercises and keep your eyes moving. Learn to glance and take in what you see without staring.

a. If the class is sitting on the floor in a circle, each member in turn should be directed to look at every other member, letting the eyes touch for a moment, so to speak. When the eyes touch, one senses a quick flash of recognition.

• Did you feel the contact between your eyes and those of the other people? Did you sense a flash of recognition? Did you glance or stare? Did you blink your eyes?

b. Without moving your head, look to the right as far as you can. Blink, then glance up and down. Next, slowly look to the left, blink, and again glance up and down. Repeat the exercise two times, trying to be aware of any tendency to become frozen.

• Were you breathing during the exercise? Did you find it difficult to blink or move your eyes?

c. Without moving your head, roll your eyes by first looking to the right, then upward, then to the left, and then downward. Make several circles with the eyes slowly. Repeat the exercise, moving the eyes in the opposite direction.

• Were you holding your breath during the exercise? Did you forget to blink? Did this exercise make you feel dizzy? Did it give you a slight headache? Did you feel any tension in the eye muscles or in the back of the neck at the base of the skull?

Exercise 62 / Face exercises

These exercises are designed to loosen the muscles of the face and thereby remove the mask so many people wear unconsciously. The aim is to restore the full range of expressivity to the face.

a. Stick the jaw out, show the teeth, and make an angry expression. As you do this, let out some sound.

b. With the jaw out, move it up and down strongly several times. Again, make a sound with the movement.

c. With the jaw forward, move it to the left and right slowly as far as it will go.

d. Stick your tongue out in a point and make an appropriate sound to express your disdain of the other person. Do it several times.

e. Wrinkle your nose up and down.

f. Raise and lower your eyebrows.

g. Keeping your jaw back and soft, reach your lips out as a baby would toward the breast. Your mouth should be open. Do this a number of times slowly.

Fig. 39. Exercising the lips

• Were you holding your breath while doing these exercises?

• Did you feel any vibration in the lips or any trembling in the jaw? These are positive responses to the exercises.

• Do you feel any tingling in your face? This is first experienced about the mouth.

• Does your face feel looser? One result of these exercises is to keep your skin healthier and more alive.

Lying-Down Exercises

The lying-down position is important in doing exercises because it eliminates the stress of gravity. In addition, this position has a regressive aspect in that it suggests a return to an infantile body attitude, which facilitates the "letting go" of control.

The basic position is lying on the back with the arms along the sides and the knees bent so that the feet are flat on the floor about 18″ apart. The floor should be covered with a thick rug or a foam pad. Let your head turn backward as far as it will go so that you are not watching yourself.

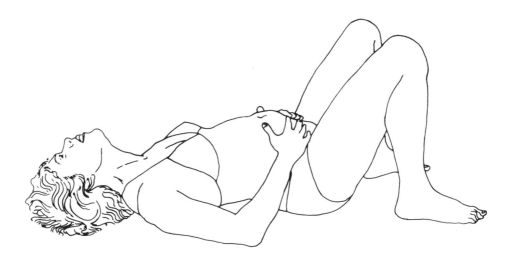

Fig. 40. Basic breathing

Exercise 63 / Basic breathing

Let your belly out as far as it will go and try to breathe abdominally.

Place both hands lightly on your abdomen and sense the rise and fall of the abdominal wall as you breathe in and out.

Do this for one minute, but do not force the breathing.

• Did your belly move upward with inspiration and down with expiration? If this didn't happen, you were not breathing abdominally.
• Did you feel any tightness in your throat? chest? diaphragm?
• Were your legs vibrating? This can happen if you do this exercise after all the others. Your legs will be highly charged then.

Exercise 64 / Leg vibrations

This is a simple exercise to get your legs into vibration. The vibration is induced by a stretch and relaxation of the hamstring muscles.

Lying on your back, raise both legs to an upright position but do not lock the knees.

Keeping the knees slightly bent, flex the ankles and push upward with the heels.

If necessary, bend and straighten the knees without locking them to get the vibrations started.

Breathe easily and let the legs vibrate for about one minute. See fig. 41 on next page.

Exercise 64-A / Variation

Do the same as the preceding exercise, but hold the toes of each foot with a hand.

Exercise 65 / Loosening the ankles

With the legs raised, as in the preceding exercise, flex and extend the foot a number of times.

Rotate the feet in a circular movement, turning them toward each other to start the rotation, and make a number of circles.

Now reverse the direction of the movement and make several circles.

Place the feet flat on the floor. Extend one leg. Shake the ankle vigorously. Repeat with the other leg.

Exercise 66 / Arching the back

Place a rolled-up blanket under the small of the back while you are lying down with your knees bent,

Fig. 41. Leg vibration. See Fig. 10.

feet flat on the floor. See if you can keep your buttocks touching the floor.

Breathe easily deep into the belly.

Let the head lie back.

Hold the position until it becomes painful.

• Did you feel any pain in your lower back at the start of the exercise? If you did, it is a sign of lower back tension, which is very common in many people.

• Were you able to keep your buttocks on the floor? If your lower back is tight and rigid, you will be unable to do this.

• Were you able to relax into the pain? If you can, you will find that the pain will ease off.

Fig. 42. Lying over roll

Exercise 67 / Reversing the arch

Place the rolled-up blanket under the buttocks and bring your knees up to your chest.

Place your arms about your knees and let your lower back curve forward.

Notice that this position feels very relaxed after the preceding one.

Exercise 68 / Bouncing the pelvis

This exercise is done from the lying-down position with knees bent

Fig. 43. Reverse arch

and feet flat on the floor. Place a folded blanket under your buttocks.

Raise the pelvis above the blanket and bounce it down onto the blanket with sufficient strength to provide a slight jar to the body.

Do this a number of times to shake the pelvis loose. This exercise can also be done using your voice as an expressive exercise. This will be described in the following chapter.

Exercise 69 / Stretching the insides of the thighs

The position is the same as the preceding exercise, with the buttocks resting on the rolled-up blanket.

Extend both legs to the sides as far as you can go.

Hold onto the blanket with your hands.

Thrust the heels upward.

9

The Expressive Exercises

These exercises are designed to help a person express his feelings, whereas the standard exercises concentrate on getting in touch with the body and relaxing its tensions without an accompanying emotional release. Inhibiting the expression of feeling leads to a loss of feeling, and the loss of feeling is a loss of aliveness. Feelings are the life of the body just as thoughts are the life of the mind.

Children suppress much of their feeling in order to adapt to their home situations. They begin by holding in the expression of fear, anger, sadness, and joy, because they think their parents cannot cope with these feelings. As a result they become either submissive or rebellious; neither of these attitudes represents a genuine expression of feeling. Rebellion is often a cover-up for need, submission often a denial of anger and fear.

Feelings arise as spontaneous impulses or movements from the core of the individual. To suppress a feeling, one has to dampen or restrict the aliveness or motility of the body. Thus, the effort of suppressing one feeling is to diminish all feeling. Yet as long as there is life in the body, there is potential for feeling.

Working with the expression of feeling in therapy, or in an exercise class, or at home helps a person get in touch with some of the suppressed feelings in his personality. The important question is: can he handle these feelings as they arise through the body work? There is some danger here, but there is also an inherent safeguard: most people will not allow more feeling to develop than they can handle. Being in therapy is another safeguard, since a competent therapist will help a person contain and handle feelings that are new and frightening. He or she can help a person understand where the feeling came from and so prevent its being acted out.

A feeling is acted out when it arises in one context but is expressed in another. For example, if a man is humiliated by a superior at work and cannot or dares not express his resentment, he may come home and beat his children. Feelings that arose in childhood situations are often acted out in adult life to the detriment of all persons involved. A woman who resented her father's indifference to her as a child may take it out, as we say, on her husband.

Expressing a feeling in the controlled setting of a therapy session or an exercise class often discharges enough of the excitation so that the feeling can be kept within proper bounds. Here is an example of how this can also be done at home. Many homemakers find the frustrations and disappointments of their daily lives intolerable. They often vent this anger upon their children, who are neither the responsible agents nor appropriate objects for these feelings. Those mothers who have been in bioenergetic therapy have found that going to a bedroom and beating the bed with a tennis racket will discharge the annoyance or anger without hurting any innocent persons. Once the feeling is discharged, their behavior becomes more reasonable. In this way they avoid becoming extremely tense by trying to contain their anger. This is standard bioenergetic practice. One of the best bits of advice that bioenergetics offers is the counsel to beat it out on a bed with a tennis racket or your fists rather than sit on the feeling or scream at your children.

Exercise 70 / **Kicking out from the hip**

This is also done from the same position.

The knees are brought up to the chest.

Holding the blanket with your hands, kick forward strongly with the left heel.

Bring that knee back and kick with the right heel.

Repeat a number of times alternately with each leg. The kick should be forward in the line of the body, not upward.

• Was the movement coming from the hip and not the knee? To get the necessary hip movement, bring your knee all the way back to the chest.

• Did you feel like saying "Get away"? When these words are used, it becomes an expressive exercise.

Exercise 71 / **Reaching out**

The position for this exercise is the same as that in the first exercise of this section, the breathing exercise.

Reach both arms upward as if you

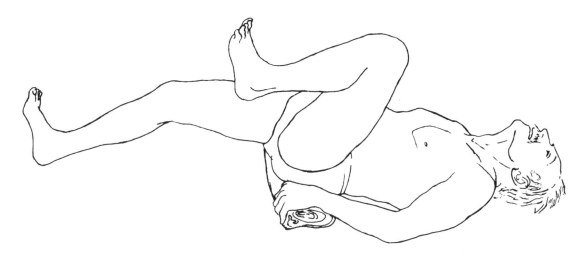

Fig. 44. Kicking over roll

106

were a baby reaching for its mother.

Make an effort to extend the reach with each expiration.

• Can you feel yourself holding back?

• Do your hands droop as in a gesture of futility?

• Do you feel yourself reaching out?

Exercise 72 / Reaching out with your lips

Let your arms lie at your sides and reach out with your lips as if to suck, as in exercise 62 in the preceding section. See figure 39 on p. 97.

Keep the mouth open and the jaws hanging loosely.

Retract the lips and extend them again in the reaching out movement.

• When the lips move forward, does your jaw come forward in an expression of defiance?

• Can you keep your breathing deep and easy while reaching out with your lips?

• Do your lips vibrate or tingle?

• Does the exercise evoke a feeling of longing?

Many of the standard exercises in chapter 8 can function as expressive exercises if you add an appropriate vocal sentiment to your movements. Reaching out, for example, becomes expressive (and can become very emotional) if one says "mamma" or "daddy" while doing the reaching. In the introductory workshops we have done, often with large groups of thirty or more, the reaching out for a mother or father will often cause many individuals to cry. Most people in our society have a considerable suppressed longing for closeness to one or both parents, which they were not able to express as children. If the longing is not too strongly suppressed, mobilizing the body through the other exercises and charging it through breathing will often bring this feeling to the surface.

We have also mentioned how the kicking-out exercise can become an occasion for self-expression by saying, "Get away." If one can awaken a feeling while doing an exercise and then express that feeling in words, the whole procedure can become quite intense and charged.

Exercise 73 / **Kicking the bed**

Do this exercise on a bed without a footboard or else on a mattress or foam rubber mat placed on the floor. Lie down with your legs extended. The kicking is done by alternately raising each leg and bringing it down hard on the mattress. The whole leg should make contact with the mattress, not just the heel. Keep your legs fairly straight while doing this exercise, but not stiff or rigid.

Kick the bed with each leg alternately in a rhythmic manner: one leg

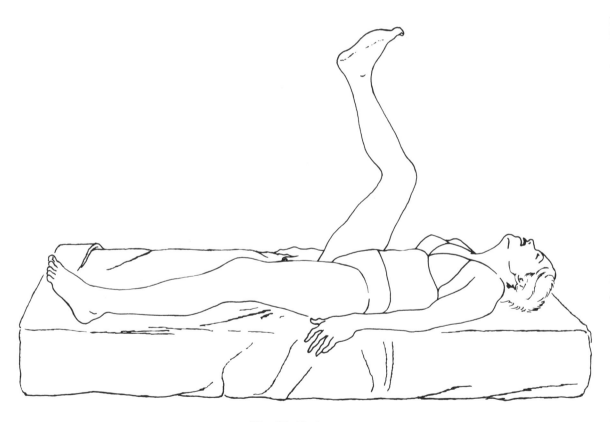

Fig. 45. Kicking

goes up as the other comes down.

Try to make the movement come from the hip rather than the knee. Do this by raising the leg as much as possible before starting the kick without bending the knee.

Say "no" with each kick in a loud, determined voice.

Now utter a loud, *sustained* "no" while you execute a number of kicks strongly.

• Did your kicks seem effective? Or did you feel impotent?

• Did your voice have a ring of conviction or did it sound hesitant or frightened?

• Could you sustain the kicking strongly or did it peter out after an initial gesture?

• Was there a coordinated rhythm between the vocal expression and the movement?

Exercise 73-A / Variation

Now use the word "Why?" instead of "no." This expression has more significance for many people than "no," probably because as children they were told that they had no right to question the dictates of their parents.

Try to prolong the sound of "Why?" as you do the kicking. You may find it spontaneously rising to a scream, in which case the feeling will have reached a climax. You will feel relaxed and relieved afterward.

Exercise 74 / Rhythmic kicking

This exercise is only partially an expressive exercise because it is done without vocal expression. It is designed to give you the sense that your legs can serve as organs of self-expression. This exercise helps strengthen the legs and charge them. It also improves your breathing.

Do the rhythmic kicking exactly as in the preceding exercise, counting the kicks. Each leg counts as one.

Do as many kicks as you can, rhythmically. Let us say that sixty is your limit. (That is low.) Do the exercise twice.

The next day, do the same exercise, adding ten more kicks to your effort. You will have to pace yourself on this one. If ten is too much, add only five.

Each day or every other day, try to increase the number of kicks. If you can reach 200, that is sufficient.

Older people have more difficulty with this exercise because their legs are tighter and their muscles less flexible. The aging process seems to hit the legs first. By doing this exercise regularly, you can help your legs stay softer and more alive. It is one of the exercises recommended for patients in bioenergetic therapy to do at home, and it will be included in the section on home exercises in chapter 13.

Exercise 75 / Banging the arms

This exercise is very easy to do. Lie on the mattress with knees bent and feet flat.

Make two fists with your hands and raise them over your head.

Bang both fists down alongside the body and say "no" with each blow.

Repeat the maneuver a number of times.

- Did the blows feel effective? Was your "no" convincing?
- Did you feel you have a right to say "no"?

Exercise 75-A / Variation

Repeat the same movements as in exercise 75, but say "I won't" instead of "no." This is a stronger expression with more ego (I) in it.

Exercise 76 / A temper tantrum

This exercise should not be done alone. Neither should it be used in class unless the participants have had considerable experience in bioenergetic work. We use it in therapy sessions where it serves to help a person let go. It is both emotionally and physically powerful.

We are including it here to give the reader an idea of the range of expressive exercises. It is not a dangerous exercise, but it can lead to dizziness if you are tightly controlled.

Lie on a mattress on the bed or on a mat on the floor. Bend the knees so that the feet are flat.

Start drumming each foot against the mattress alternately with bent knees.

Bring the knees well back on the body so that you are using a hip movement rather than a knee movement.

Do this for a while, then stop.

Start the drumming of the feet again, and as you do it, pound your fists alternately into the mattress. You are now using your legs and arms.

Repeat the above procedure, and let your head turn left and right with the movement of your body.

In a loud and sustained voice, shout or scream "I won't" as you do the movement.

The key to this exercise is the coordination between the movements of the legs, arms, and head. When the exercise is properly executed, the body moves as a unit. The left leg and the left arm move together; that is, both strike the mattress at the same time. The head turns toward the side of the blow rather than away from it. If the left arm moves together with the right leg, it is as if the person were at cross purposes. Correctly done, the body moves like a spinning top. It is beautiful to watch. Strangely, you will not feel any dizziness in this situation if you do the exercise properly. The dizziness develops when you are not going fully and freely with the movement; there is an unconscious holding against the expression.

Exercise 77 / Reaching out with lips and arms

This exercise allows a person to experience the longing of an infant for its mother. It can become very emotional when the word "mamma" is used and the longing is evoked.

Lie on the floor with knees bent and feet flat. Breathe easily and

Fig. 46. Temper tantrum

deeply into the belly.

Reach up with your arms and reach forward with your lips as in figure 39 (see p. 97).

Say "mamma" and see how much feeling you can put into your voice and arms.

- Did you sense any holding back of the feeling?
- Did it embarrass you? Did you feel silly reaching out like a child? Remember that we are all children at heart and that we had two parents to whom we will always remain attached in our hearts.
- Did the reaching out bring up a feeling of sadness? If so, can you cry or does your throat feel choked?

Exercise 78 / Demanding

Instead of reaching out as in the preceding exercise, make two fists and raise them in front of you.

Shake both fists strongly.

Say "Why?" "Why weren't you there?" "Why didn't you care?" You can use any other expression that seems appropriate.

- Could you get any emotion into the expression? If you couldn't, it indicates that you are holding back either because the situation is inap-

propriate or else because you are inhibited.

Exercise 79 / Expressing anger

A person should be free enough to be able to express his anger physically where it is appropriate. Most people are too frightened of violence to be *able* to express anger in a physical way without extreme provocation. There is a taboo in our culture against hitting, which is unfortunate since it mostly operates to render innocent people helpless before bullies.

Stand in front of a bed. It is best if the bed has a foam rubber mattress, so that you cannot hurt either yourself or the bed. This is an indispensable exercise if you suffer from tensions in the shoulder girdle because these tensions relate largely to inhibitions in the use of the arms to strike with. There are several variations to the exercise.

Stand with your feet about 18'' apart and bend both knees slightly.

Make two fists and raise them above your head.

Raise your elbows and pull them as far back as you can.

Now hit the bed with both fists strongly, yet in a relaxed way without forcing the movement.

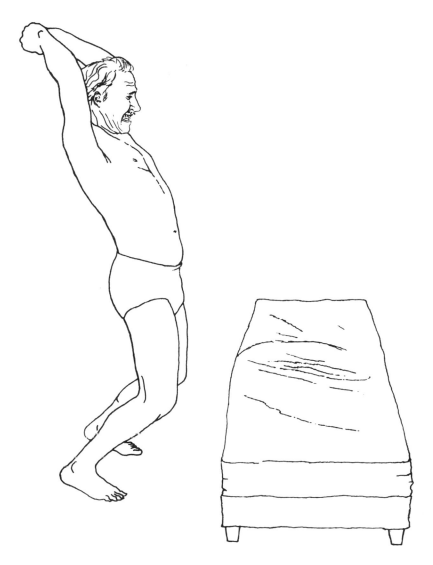

Fig. 47. Hitting with fists

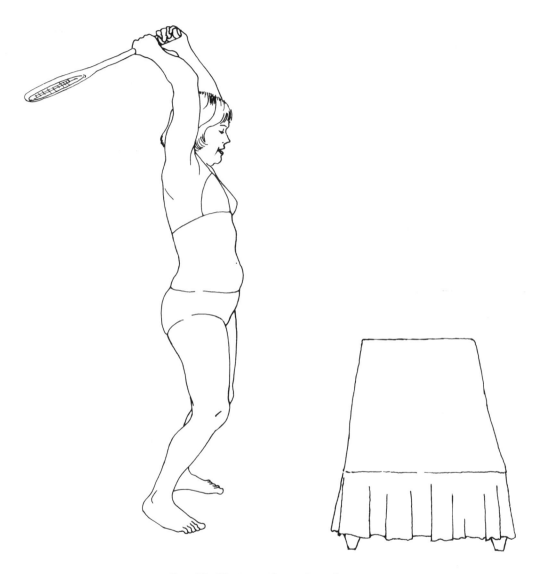

Fig. 48. Hitting with tennis racket

114

Say any words that express a feeling of anger. You can use words such as "No!" "I won't!" "Leave me alone" "Damn you!" or "I hate you!"

• Can you sense your blows as being effective or do they feel impotent?

• Can you feel any emotion with the exercise? It is not necessary to have a strong feeling to make this exercise meaningful.

• Are you frightened of your own potential for violence? In that case, repeated use of the exercise will reduce your anxiety and give you more control over your anger.

Exercise 80 / Using a racket to express anger

The same exercise is done using a tennis racket instead of the fists. The racket adds a sense of power and helps overcome a sense of impotence.

Raise the racket over your head and strike repeated blows with the flat surface of the racket against the bed.

Say whatever words express any feeling you have. Remember, you are not hurting anyone with this exercise.

• Did you have a feeling of satisfaction from your blows?

• Did the sound of the blow frighten you? Did it feel too violent? Were you afraid that you might want to kill someone? If you can accept and express this feeling ("I'll kill you") while hitting the bed, you will discharge some of your murderous rage and gain control over the feeling. There will be less likelihood, then, of your releasing that rage in a real-life situation.

Exercise 81 / Rhythmic hitting

This is an exercise that we have used personally and that many patients use regularly. It is like rhythmic kicking and serves to strengthen the arms, develop coordination in arm movement, and release tension in the shoulder girdle.

Raise both fists, or the tennis racket, and deliver twenty blows to the bed.

The hitting should be rhythmic, neither too slow nor too fast.

Breathe in as your arms go back, then breathe out fully as they come down.

Stretch your arms fully behind your head to mobilize the maximum energy for the blow. Note that a well-

115

Fig. 49. Twisting a towel

known law of muscle action is: the greater your stretch, the more effective and powerful will be your resulting contraction. You do not have to hit hard. The power of the blow comes from the stretch and the timing. Think of shooting an arrow. Similarly, one doesn't have to push a blow. If you make a full stretch, the blow will develop spontaneously as you let yourself go. After you can deliver twenty blows without strain, increase the number regularly until you can do forty or fifty blows.

Exercise 81-A / Variation

A different kind of hitting can be done by striking the bed with one arm after the other. This exercise involves a stretch of the body that uses muscles not reached by the preceding exercise. Again, the important things are:

Keep loose and fully stretch each arm before the blow.

Breath in on the full stretch and breath out with the release.

Try to bring each fist back over the ear of the other side of the body to get the maximum tension for the blow.

Exercise 82 / Aggression

The word itself means "to move toward." People use it as denoting "going for what one wants." An aggressive person is one who moves toward the fulfillment of his wants. The lack of aggression means passivity—waiting, not reaching out.

A good exercise in aggression is the twisting of a towel. Take a medium-sized turkish towel and roll it up. Then twist it as strongly as you can with both hands.

As you twist the towel, say "Give it to me." Keep twisting the towel and saying "Give it to me."

• Did you reach the point where you feel that you can get what you want?

• Did you loosen your grip after each demand or were you able to hold on?

• Did your voice sound strong and sure?

• Do you have the feeling that you can get what you want? Wouldn't it be a nice feeling to have?

Every expressive action is an aggressive act in the sense that it is a "moving out" toward the world with one's feelings or energy. We should not think of aggression only in the negative terms of political science. Reaching out to love or for love is an aggressive action. Saying "I love you" is as aggressive as saying "I hate you." The aggression is in the action of reaching or saying, not in the content of the words. On the other hand, to have a feeling one cannot express is a sign of passivity. Every act of self-expression involves some degree of aggression.

To be fully self-expressive, the body has to be free from its tensions, specifically those tensions that block our natural aggressiveness. Because our aggression has been blocked since childhood, it requires considerable work to free it. The repeated use of these exercises can be of considerable help in this endeavor.

10

Working with
the Bioenergetic Stool

Lying over a bioenergetic stool is an important part of bioenergetic body work. It helps to stretch the tight back muscles that are difficult to get at otherwise. It helps you breathe deeper without making too great a conscious effort. If you lie over the stool and relax into its stress, your breathing deepens spontaneously. You don't have to work at it. Similarly, the position itself stretches and releases the tense back muscles.

The current bioenergetic stool is an adaptation of the wooden kitchen step stool originally used. One or two tightly rolled-up blankets are strapped to the stool. We use army blankets because they provide a firmer roll. The stool itself is 24'' high, and the rolled-up blankets generally add another 6'' to 8''. Like its original, the kitchen step stool, the legs are flared and braced with crosspieces to provide a wide and solid base. Figure 50 shows what such a stool looks like. A 1'' wooden dowel is attached to the underpart of the platform (in the illustration it is inserted through the holes shown) and extends 5'' to 6'' on either side. This provides a handgrip to help you rise up from the stool. Bioenergetic stools are available from a number of manufacturers, the price varying according to the wood and workmanship. A list of places where bioenergetic stools may be purchased appears in the Appendix.

There are many ways of working with the stool. The primary exercise is to lie with your back on the stool, with the blanket at the level of the lower ends of the shoulder blades. This is the same level as a line joining the two nipples. This level is close to where the main bronchus (air duct) divides into two branches, one going to each lung. It is an area of severe constriction in most people. Lying on the stool, the person reaches his hands back to a chair that is just behind the stool.

Exercise 83 / Lying over the stool

To go over the stool easily, stand with your back toward it and put both hands on the blanket roll behind you. Then slowly lower your back until it rests on the roll and let go with your hands. The stool will support your weight. Now raise your arms and reach backward to the chair. Bend your knees, keeping your feet flat on the floor.

Lie on the stool as long as you reasonably can, but no longer than one minute the first time. Try to sense what is going on in your body.

When you come up, do not get off the stool immediately. Lift your head and place your hands behind it for support as in figure 52, p. 121. This is the resting position which allows

Fig. 50. Bioenergetic stool

119

deeper breathing to continue without stress.

• Were you able to reach the chair back behind you? Did you feel any pain in your upper back where it rested on the stool?

• If your back was very rigid and tight, you may have been unable to touch the chair and the stretch can be quite painful. Come up and try again. The pain generally disappears with practice, as the muscles of your back relax. In time, the exercise even

Fig. 51. Leaning back on stool

becomes pleasurable.

• If you have bursitis in one shoulder, you may find it impossible to stretch that arm backward to touch the chair. Do not force it. Do the exercise with only one arm reaching backward. However, in all cases that we have worked with, the shoulder exercises described in chapter 8 together with the continued use of the stool resulted in a marked alleviation of the bursitis.

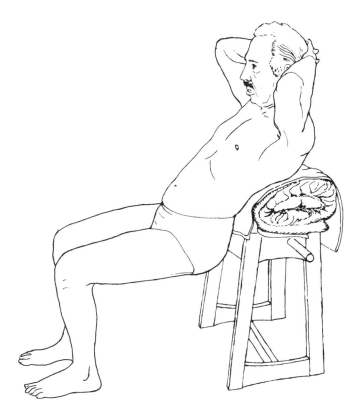

Fig. 52. Resting position

• Did you feel any tension in the diaphragm or abdominal muscles? When the abdomen is tightly contracted, the stretch of the abdominal muscles in this position can be somewhat painful. This pain disappears as these muscles relax with deep breathing.

• Did you feel any pain in your lower back? Again, this is a sign of considerable tension in that area.

• Were you able to feel your feet on the floor? If you were, you may have become aware of some charge or excitement flowing into them manifested by tingling or other paresthesias (pins-and needles sensation). Did you experience any tingling in your arms or face?

• Did you have trouble breathing? Did you feel your throat tightening or becoming choked? This is a sign that you are unconsciously holding against deep breathing. You can overcome this problem to some extent by making a sound as you breathe out. The feeling of choking can also be caused by blocking an impulse to cry, that is, choking it off. If you sense this impulse, try to express it.

Rock back and forth repeatedly, each time trying to stretch the arms further back.

If you sense that your muscles have relaxed somewhat, try to hold the back of the chair for thirty seconds or one minute.

Come up into the resting position with your hands behind your head and breathe easily for a while.

When you decide to come off the stool (you should not be on it more than two minutes at a time), place your hands on the blanket or on the handles and push yourself up onto your feet.

Exercise 83-A / Variation

If the position described in exercise 83 is difficult for you to attain or to maintain for more than a few seconds, there is a rocking exercise we use to reduce the stress.

With your back on the stool, hold your arms upward. Reach back with them as far as you can, then bring them upward. Let your head follow the movement of your arms. This way, you place only a momentary stress on the back and shoulders rather than experiencing the continual stress of the first exercise.

Repeat Exercise 1 / Basic vibratory and grounding exercise

After being on the stool in any position, the person should reverse the backward arch by bending forward. This is done by doing the first exercise, page 11, figure 2, called the forward bend. You may have noticed that in all bioenergetic exercises a movement in one direction is followed by a movement in the opposite. Such a procedure increases the flexibility of the body and, by extension, the flexibility of the personality.

Exercise 84 / Different positions on the stool

The bioenergetic stool is used to stretch and relax all the muscles of the back. In exercises 83 and 83-A the pressure was focused on the upper back. But the pressure can also be directed at other areas of the back simply by moving up or down on the stool. If you lie on the stool with the blanket roll in the middle of your back, the stress will affect the muscles of this area as well as the diaphragm, which has its insertion in the middle vertebrae. The stool can be placed alongside a bed to give you a sense of security when you lean backwards.

Place the stool alongside a bed as in figure 53.

Put your hands on the stool behind you and let the middle of your back rest on the blanket roll.

If the stress seems too great, come up to the resting position described above. Then try to go back over the stool again. Since tension is an expression of fear, you will find the exercise easier to do as you familiarize yourself with it.

If you can, let your arms go over your head and touch the bed behind you.

Stay in this position up to a minute, breathing and sensing your body.

Come back up to the resting position and hold it for about thirty seconds.

• Was this stress too much for you? Could you feel the tension in your middle back?

• Did you have difficulty breathing in this position? Could you feel the tension in your diaphragm? When you come off the stool, bend forward as in exercise 1, the basic vibratory exercise (see p. 11), and let your legs vibrate for a while.

Exercise 85 / Lower back stretch

In the beginning, most people find this to be a strenuous exercise. With practice, it becomes somewhat easier. We use it regularly in therapy sessions to relax the lower back muscles and to open the pelvis.

The stool should be alongside the bed. If the bed is low, place a pillow on the bed to rest your head on. Stand with your back to the stool, place your hands on the blanket roll, and put your lower back there.

Lean back over the stool and let your head rest on the bed or pillow.

Keep your hands loosely on the handle of the stool until you feel relaxed. Try to keep your feet flat on the floor.

Fig. 53. Lower back on stool

Let yourself go to the pain in your lower back and breathe easily and deeply.

Try to let your pelvis drop. Do not stay in this position more than a minute, and come up into the resting position when the pain is too strong.

• Were you able to relax in this stress position for thirty seconds? Your ability to tolerate this stress depends on how free from tension your lower back is.

• Did you feel as if your back would break? That feeling represents an intense fear.

• Could you breathe into your pelvis? The more you can relax in this position, the deeper your breathing will go.

• Could you let your pelvis drop and keep your feet on the floor?

Repeat the forward bend you did after the preceding exercise in order to get the vibrations going in your legs again. You may now have strong vibrations throughout the pelvic area.

Exercise 86 / Pelvic stretch

In this exercise, place the buttocks on the stool and arch over backward. The stool should be alongside a bed and your head should rest on the bed. When on the stool, hold onto the blanket or the handles with both hands. Your feet will be off the floor.

Let your feet hang, pressing down with the heels.

Stay in this position for up to one minute, breathing easily.

Holding the handles firmly, bring both feet straight up in the air and press upward with your heels by flexing the ankle. Your legs should go into nice vibrations in this position.

To come off the stool, swing your legs downward strongly while holding the handles. This will bring the upper part of your body off the stool and your feet will touch the floor. See figure 54 on the next page.

• Did you feel any tingling in your legs in this position? Did you experience a sensation of pins and needles in your feet?

• Did this exercise open up more sensation in your pelvis?

• Could you feel the tightness of your buttocks?

• Did your legs vibrate strongly when they were extended upward?

Exercise 87 / Kicking over the stool

This is a variation of the preceding exercise that aims at getting a stronger charge into the legs while the pelvis is extended. It is done from the position of the previous exercise while you are lying on the stool on your buttocks.

Hold onto the handles of the stool, bring one knee back up and kick forward strongly with the heel. Try to direct the movement downward.

As you kick with one leg, bring the other knee up for the next kick.

Now kick alternately with each leg, using some force.

• Did you feel the stretch of the thigh muscles as you kicked?

• Did you feel a stretch of the hip joint?

• It is advisable to follow up this exercise by bending forward to get the feet in contact with the floor and to have the legs vibrate.

Fig. 54. Ass on stool (pelvic stretch)

Exercise 88 / Chest pressure

In this exercise, pressure is placed on the chest, thus mobilizing that area and facilitating the breathing.

Place the stool in an open space and lie over it with your chest on the blanket roll.

Let your head drop forward. Let your arms dangle.

• Did you sense how tight your chest is? Did you have any difficulty breathing?
• Could you feel your chest relax under this pressure?

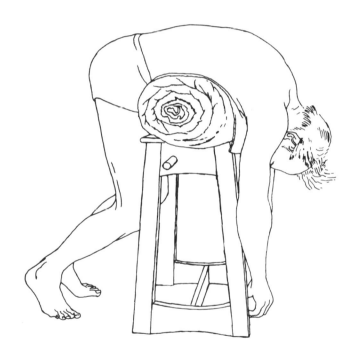

Fig. 55. Chest on stool

The great value of the stool exercises is that they facilitate breathing without effort. The only exception is exercise 87, kicking over the stool. One has to relax or give in to the stretch or stress, not fight it. As one learns to do this, the breathing will become deeper as a natural response to the stress. This may take considerable practice, as the stress is quite strong for some people. Do not regard these exercises as a challenge to your will or endurance. Do not persist in them when they become painful. The pain will cause you to tense up. You can always try again. Eventually, you will feel more comfortable with the stool work, but it never becomes easy.

If you can relax on the stool, it will become like a meditative experience with your body. You will sense your breathing as a spontaneous function of your body, and you may become aware of other involuntary movements. When this happens, the stool is working for you.

11

Sexual Exercises

These exercises are designed to bring more feeling into the pelvic area and to loosen the pelvis for greater pleasure in sex. They do not involve the genital organs nor do they induce any genital excitation. The amount of sexual pleasure, that is, orgastic pleasure, you experience depends on how much sexual excitation you can allow to build up in the pelvis before the discharge. In a sense, then, the pelvis functions like a condenser. Its capacity is determined by its inner largeness and its mobility. Muscular tensions within the pelvis limit its capacity, while tensions in the external muscles reduce its ability to discharge the excitation.

The pelvis should be loose so it can swing freely as the current of excitation passes through the body. Deep abdominal breathing is the critical factor in enlarging the pelvic reservoir. Deep breathing is therefore a primary objective in all exercises, especially the sexual ones. In addition, through these sexual exercises we seek to release the tensions that inhibit pelvic mobility and to integrate the pelvic movements with the respiratory waves.

The best results will be obtained from these exercises if you do them in conjunction with the other exercises in this manual. The body is a unity: tension in any part interferes with and restricts the natural movements of all the other parts. This is particularly true of tension in the neck. If the head is held rigid by tense neck muscles because of fear of "losing one's head," the pelvis cannot move freely. In these exercises, keep your head back and let it go; it will not fall off. If you should become dizzy or anxious, stop the exercise, assume a resting position, and allow your breathing to quiet down. Then try the exercise again, but do not push yourself against your fear or pain.

Exercise 89 / Pelvic swing or bounce

This exercise involves the use of the stool. It is introduced here as a sexual exercise because it focuses mainly on the movement of the pelvis.

Lie on the stool with the blanket roll against the upper part of your back.

Reach back with both hands to a chair behind the stool. It should be a fairly heavy chair.

Keep your feet flat on the floor while doing this exercise. Many people let the feet leave the floor and so lose contact with the ground.

At the same time, try to maintain your grip on the chair. In that way, you are anchored at both ends.

Bounce the pelvis up and down rhythmically. Moving rhythmically is important in this exercise. Start slowly, then move faster as your body loosens up.

Try to keep breathing in time with the pelvic movements.

• Could you get your pelvis swinging in this position?

• Did you experience any pain or immobility in your lower back? Such pain or immobility inhibits the natural swing.

• Could you keep a rhythm going? Was your breathing coordinated with the pelvic movements?

• Did you keep your buttocks loose as your pelvis swung upward?

• Did you keep your heels on the floor?

The important element in this exercise is to remain anchored in your grip on the chair and in your feet. If the two ends of the bow of your body are securely anchored, the movement will be correct. This is not an easy exercise to do correctly. The natural pelvic swing is inhibited in most people, and they tend to lift the pelvis upward rather than letting it swing. Another tendency is to tighten the ass, which will block any sexual feeling. As you work with this exercise, try to keep your ass loose.

Follow this exercise by bending forward in the grounding position, letting the vibrations develop in the legs.

Exercise 90 / Stretching and relaxing the muscles of the inner thigh

These muscles (adductors) act to bring the thighs together. They were formerly called the morality muscles because most girls were taught to sit with their legs pressed together.

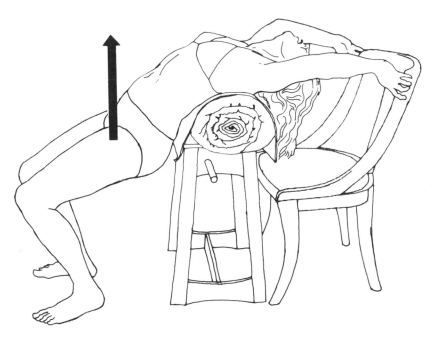

Fig. 56. Bouncing pelvis

These muscles are quite contracted in many people (both men and women). Stretching them helps to relax the pelvic floor.

Lie on your back on the floor with a rolled-up blanket placed under the small of your back. Your ass should rest on the floor. If this seems too difficult, reduce the size of the roll.

Bend your knees and bring the soles of your feet against each other.

Let your arms lie at your side or rest lightly on your inner thigh (the adductor muscles). Let your head turn back as far as it will go.

Press downward with your ass against the floor, and spread your knees apart, keeping the soles of your feet in contact with each other.

Hold this position for several minutes while breathing deeply into your abdomen. Your belly should be out.

Remove the blanket roll if your lower back becomes painful. In that case, you can do the same exercise without the blanket roll, though it is less effective.

• Did you feel any pull or stretch in the adductor muscles? Did they become shaky or begin to vibrate?

• Could you spread your knees widely while keeping your ass pressed against the floor?

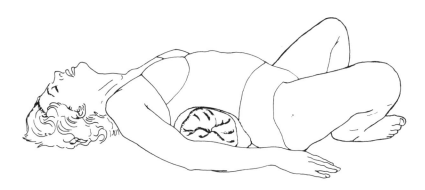

Fig. 57. Adductor stretch

• Could you breathe into your belly?

• Could you keep your ass loose and your anus open during the exercise?

When you finish the exercise, remove the blanket roll and bring your knees up so that your feet are flat on the floor. You will then be ready for the next exercise.

Exercise 91 / Vibrating the inner thigh muscles

From the above position, place your feet flat on the floor about 18'' to 24'' apart. Move your knees slowly apart without moving your feet. Try to keep your feet flat while doing this. The movement should be very easy and effortless. Spread your knees as far apart as they will go, keeping your feet fairly flat.

Slowly and easily, bring your knees together until they touch. It is important to move slowly and easily to induce the vibrations to occur.

Now move your knees apart again in the same easy way and then back again.

When you sense the vibrations beginning, continue with the move-

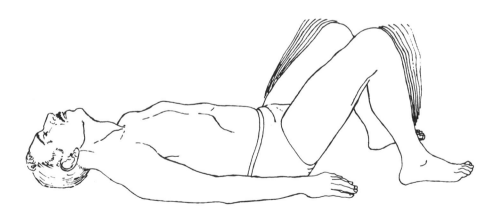

Fig. 58. Adductor vibration

ments so that the vibrations will continue and grow stronger. You will find these vibrations very pleasurable. Remember to breathe.

• Could you get your legs to vibrate in this position?

• Did you notice how your breathing became spontaneously deeper as the vibrations developed?

• Did you experience any pleasant sensations in the thighs and pelvic floor?

Exercise 92 / The circle or full arch

This exercise provides a strong stretch of the muscles on the front of the thighs; it can be painful if these muscles are tight. This muscle tension prevents the pelvis from swinging freely. In the exercise, the pelvis is suspended between the feet and shoulders, which allows it to vibrate freely if you can relax in the position.

Lie on your back on a mat or bed, bend your knees, and keep your feet flat about 12″ apart.

Take hold of your two ankles with your hands, and arch over by pulling yourself forward with your hands and letting your head fall back. Only your head, shoulders, and feet should touch the bed or mat.

Push your knees as far forward as they will go.

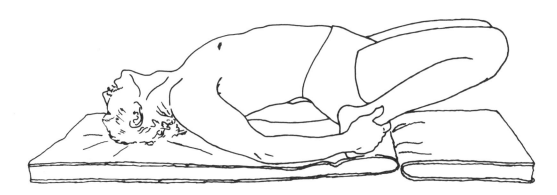

Fig. 59. Arch—full circle

Let your pelvis hang freely. Do not tighten your ass. Try to keep your anus open in this and all other exercises.

If the arch is painful, do not hold the position but let your back come down to the bed again. Then try the arch once more.

• Were you able to hold the arch and breathe into your belly while doing so?

• Did your pelvis develop any vibration in this position? Were you able to keep your ass loose?

Exercise 92-A / Variation

If you have difficulty with this position—and many people do—try to rock backward and forward on your feet while in it. When you rock forward, try to touch the bed with your knees. When you rock backward, the strain is relieved. In this way, you can slowly stretch the front thigh muscles.

Exercise 92-B / Variation

Assume the arch position as above, but place your fists under your heels.

Press downward with the heels, but do not let yourself rock back-ward. Keep your knees pointing forward and down.

Bounce the pelvis with several quick movements to set it in vibration if this does not happen spontaneously. Keep your ass loose; do not push or tighten it. See figure 60 on the next page.

• Did you develop any vibration in the pelvis?

• Were you able to breathe into your belly?

• Could you keep your ass loose?

In these exercises, the balls of the feet act as fulcrums for the lever action that will lift the pelvis. Pressure is exerted through the heels by pressing them downward. If the knees are kept forward and down, the resulting force will swing the pelvis upward without any tightening of the ass or abdomen.

Exercise 93 / Socking the ass

This position involves bringing the ass backward so that it becomes charged and mobilized for forward movement. This forward movement should come from the feet, not from the ass. Pushing forward from the ass tightens it and cuts off much of the sexual feeling.

Lie flat on your belly on a mat, bed, or floor.

Place your hands flat, then bend and spread your elbows so that your chest is against the mat. Turn your head to one side.

Dig your toes into the mat, so that you can press on them. Bend your knees slightly, and keep them against the floor.

Keeping your belly against the mat, pull your ass backward as much as you can.

Hold this position and press strongly with your toes, breathing easily and deeply.

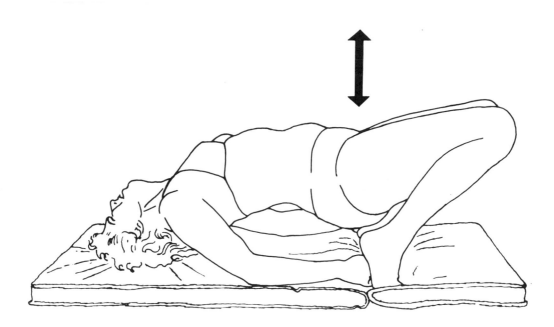

Fig. 60. Arch with fists under heels

Bounce the pelvis rapidly against the mat while pressing down with toes and knees.

- Could you feel your belly pressing against the mat? Were you able to breathe into your belly?
- Did your pelvis develop any vibrations doing this exercise?

To mobilize the full sexual charge, it is important that the legs be grounded. This can only be done if the feet are pressing against some support during the sexual act. We have recommended that the person in the upper position in the sexual act press his feet against the footboard of a bed to get the necessary grounding for the thrust. In the absence of a footboard, you have to dig into the mattress with the toes as in the above exercise.

People who have tried this exercise report that it greatly increases the amount of their sexual feeling. This exercise is one way to experience the buildup of charge for the thrust.

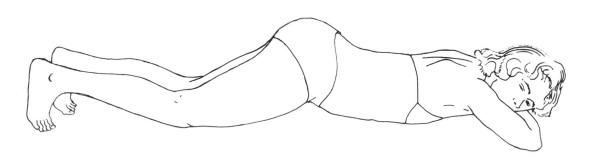

Fig. 61. Socking the pelvis

Exercise 94 / Pelvic vibration

Do this exercise in front of a bio-energetic stool, a chair, or a table. The reason for this is that it places you in a position where strong vibrations will be induced in the legs. These vibrations can be extended to the pelvis.

Stand with your back to a stool, chair, or table with feet pointing straight ahead and about 6″ apart. Place a folded blanket on the floor in front of your knees in case you let yourself fall.

Place both hands behind you, touching the stool, chair, or table lightly for balance. There should be no weight on the hands.

Bend both knees and pitch slightly forward so that the heels just come off the ground. You should have your weight on the balls of the feet.

Balancing yourself in this position by your hands on the stool, chair, or table, arch your body backward and tip the pelvis back also without breaking the arch.

Hold the position, breathing deeply, until your legs begin to vibrate.

When your legs are vibrating, move your pelvis softly forward and back. The movement should come from the legs and feet. The vibrations should start up in the pelvis, and you may experience some spontaneous pelvic swings.

If the position becomes too painful in the thighs, you should let yourself fall onto your knees. Then get up, walk around the room, and repeat the maneuver.

• Did your pelvis move spontaneously? Were you frightened by this movement?

• Did you find your knees shaking to and fro rather than vibrating up and down? This side-to-side shaking of the legs is an expression of fear.

• Do you feel your pelvis more alive now?

These exercises can help you increase the sexual charge in your pelvis. That charge is indicated by the development of spontaneous pelvic movements. However, if you cannot contain the charge, the pelvis will respond too quickly. Holding the pelvis back in any of the exercises allows the charge to build to a higher intensity before the involuntary movements of discharge take over.

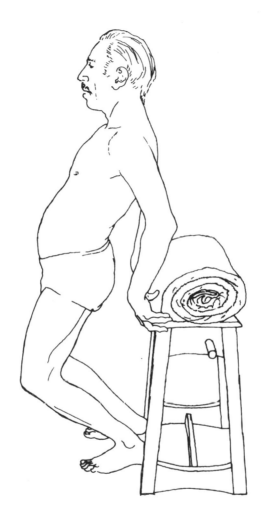

Fig. 62. Pelvic vibration

12

Massage Techniques

Massage is an important part of bioenergetics. It is a counterpart to the active exercises in which effort is required to produce results. In massage, you do nothing but relax and enjoy the touch and pressure of the masseuse's hands as they stroke and knead the skin and muscles. Sometimes, however, when an area of muscular tension is involved, massage can be painful. As I have been massaged regularly over the years, I have experienced pain in different parts of my body. Knowing that the pain was due to tension, I tried to go with it, that is, stay soft and let it hurt. The masseuse could also feel the tension in the form of tight muscles in certain areas. Generally, after several massages, the pain disappeared and the area felt soft to the touch. Fortunately, the massage was always more pleasurable than painful, and I felt so good after it was over that I looked forward to the next one. Very often, toward the end of a good massage, I would fall asleep on the table.

Massage serves several purposes. We all need to have something nice done for us and to us. Massage partly fulfills this deep oral need, which is one of its attractions. But we also have an adult need to be touched in a pleasurable way without any sexual undertones, and massage meets this need, too. Equally important is the work on the tense muscles. The hands of a masseuse can get at tensions that are inaccessible to our own hands and not directly affected by the exercises. Muscular spasticities at the base of the skull and in the floor of the mouth meet this description. Very often these muscles will relax to a firm and steady pressure.

Much of the value of massage depends on the sensitivity, skill, and hands of the person doing the massage. One has to know just how much pressure to apply. Too much will make you too tense; too little will have little effect. Since massage is a function of touching, one has to be in touch with the feelings of the person being massaged. Is he frightened? People are frightened of being touched, and they are afraid of being hurt. Is she breathing or is she holding her breath? Holding your breath alone will make the whole procedure more painful than pleasurable. The skill of a masseuse is to be able to feel the tight muscles in a person's body and to know what pressure or manipulation will relieve them. One also needs the experience of having done massage. Finally, the quality of one's touch, of one's hand, is critical. Cold, lifeless hands will make a person shrink rather than relax and expand to the touch. And weak, flabby fingers offer no stimulation.

Touching is an energetic process of contact. Through touch, energy flows from one person to the other. That is why the touch of some hands can have a healing effect. If you do any massaging, you should keep yourself relaxed and charged. Your own movements should be easy and not mechanical. You are not working with wood or stone. You should keep your breathing deep and full to have the energy to put into your hands. And you should not do any massage if you do not enjoy it, as your touch will not be a positive experience for the other person.

This is not a book on massage, which is an art in itself. And although massage cannot do what the exercises do, we use some massage techniques regularly in our exercise classes for the reasons stated above. The massage is done by the participants with each other. It helps bring people directly and literally in touch with each other and promotes a sense of contact and closeness in the group. In this chapter we describe these techniques, which can be done by persons without any prior training or experience in massage.

In addition, a certain amount of massage can be done at home. A husband can help his wife relax by working on a few critically tight areas. A wife can do the same for her husband. One can sometimes relieve a headache (a tension headache, not a migraine type) if one knows a few simple maneuvers. We shall describe how that can be done.

We also have a very fine back walk that can be done between husband and wife under some safeguards that is very relaxing.

Exercise 95 / **Massage of back and shoulders**

The person to be worked on sits on the floor cross-legged. It is best if the shoulders are bare. The person doing the massage stands or kneels behind his partner; take the most comfortable position you can. Both of you should be relaxed and breathing easily.

The following instructions are addressed to the person doing the massage.

Place both hands lightly on the shoulders of your partner and move your fingers over the shoulder and onto the back, pressing gently to feel any tense muscles. These will be felt as hard knots under your fingers.

With your fingers touching softly, use your thumbs to knead these knots. If it is too painful for your partner, do it more gently. Work over the whole upper back.

Place your hands on the shoulders and press down strongly and steadily until the person lets his shoulders drop. Do not force; a steady pressure with your weight behind it will do the job. Make sure your partner is breathing.

Using the sides of your hands, drum moderately on the back of the person, moving around on the tense areas. This will jar loose some of the tensions.

If your partner wants a stronger treatment, use your knuckles to knead the hard muscular masses around the shoulder blades. Remember that you are doing this to help your partner feel good, not to hurt him.

Now, with your fingertips, again massage the whole area, working out

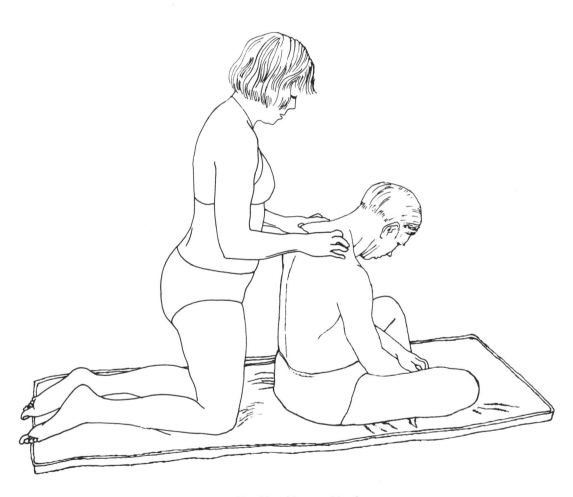

Fig. 63. Shoulders and back

over the shoulder joints into the upper arm and downward along the back. Critical areas are around and on the shoulder blades.

• Could you feel any contracted or spastic muscles? Few people are without them.

• Did your partner's shoulders drop and relax with your massage?

• Did his breathing deepen and did he enjoy your massage?

Exercise 96 / Massage of the neck muscles

The neck is an area where tension develops early in life and persists. Tight neck muscles may have several possible causes. A stiff neck denotes false pride and stubbornness. A short neck, generally due to the contraction of the muscles at the side of the neck, can be interpreted as due to a fear of "sticking one's neck out."

A thin neck indicates a lack of full communication between the head and the rest of the body—a narrowing of the channel, as it were. In all cases, neck tension expresses a need to hold onto one's head, a fear of letting go of the head. Where such tension is severe and long-standing, it can lead to an arthritic condition in the cervical vertebrae.

Massage of the neck muscles will not remove the tension, but it can help considerably. Often, quite strong maneuvers are required to get a person to let go of his head.

With your partner sitting cross-legged, kneel alongside him on his left. Place your left hand on his forehead to provide a firmer support and your right hand on his neck. In doing this massage, follow all the precautions listed above.

With the fingers of your right hand, palpate (feel) the neck muscles from the base of the skull to the root of the neck. Sense the quality of the muscles; some may be tight as steel bands, others taut as wire, and some may feel hard and knotty.

Using your fingers, knead the tight muscles while you support the head with your left hand.

It is important to tell your partner to keep breathing and to make some sounds so you can stop if it is painful.

• Can you feel the tensions at the base of the skull? Along the back and sides of the neck? At the root of the neck?

• Does your partner feel his head to be looser as a result of your work?

• Were you breathing easily while you were working?

• *Note:* Left-handed people

should work from the opposite side, since their left hands are stronger for the massage work.

Exercise 97 / Relieving a tension headache

The position is the same as the preceding exercise. Start the massage at the base of the skull and work upward over the top of the head.

Place your left hand on your partner's forehead and your right hand at the junction of head and neck.

First, feel for tension all along this area reaching to the mastoid bone behind the ears. To feel the tension, press firmly and move the fingertips.

With the first three fingers, massage the muscles strongly. There should be some slight pain if you are working correctly.

With all your fingers, slowly massage upward, moving the scalp until the two hands come together.

Now move to the other side of your partner. Place your right hand on his brow, fingers encompassing the crown. Place your left hand on the back of his head. You should have the whole scalp in your hands. With your fingers, move the scalp back and forth, freeing it up. If you can loosen the scalp, the headache will generally disappear, since it is caused, most likely, by a tension that surrounds the head just below the crown.

Although this massage technique developed as a means of removing headaches, it has become part of the regular bioenergetic massage routine. After it is done, most people feel that their heads are lighter.

If there is only partial relief, you can try the maneuver again. However, if it doesn't work, do not persist. If the headache is migraine or a tension and migraine headache, it will not yield to this procedure.

Exercise 98 / Lower back massage

This massage is done with a person lying flat on his belly on a mat, mattress, or bed. His partner gets on his knees, with one knee between the legs of the person lying down. See fig. 64 on the next page.

Place both hands on the waist of the person, with thumbs pointing toward the midline.

With your thumbs, feel the muscles in the lumbosacral area, that is, between the ribs and the buttocks.

Knead these muscles with your

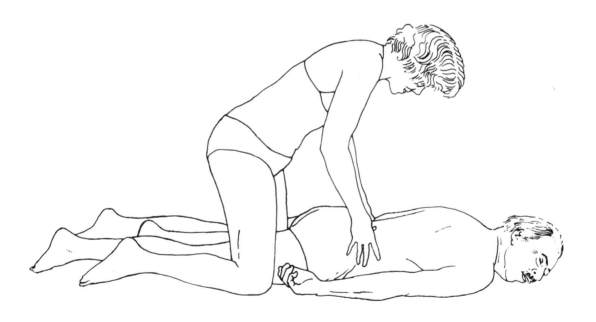

Fig. 64. Lower back

thumbs, pressing firmly and moving up and down.

Make sure that your partner is breathing deeply while you work.

You can also use your fists against these muscles. Place both fists in the small of your partner's back, and press down strongly when your partner is breathing out. Release as he breathes in, then apply the pressure again as he breathes out. This is done a number of times to loosen these very strong and often very tight muscles.

You can also work upward on the back with your fists, pressing down on the muscles alongside the spine when your partner is breathing out. The pressure will facilitate his breathing out.

This exercise should not be done if the person is in a state of acute lower back pain. It is a helpful exercise, however, if the person suffers from a low-grade chronic ache in this area.

Exercise 99 / Buttock massage

This is done from the same position as the previous exercise. The hands are placed on the buttocks, and the thumbs are used to press against and massage all the muscles in this area.

The best procedure is to work on both sides simultaneously, starting along the top of the buttocks, then downward. Next, press firmly into the central mass of the buttock muscles.

Do not cause undue pain. If it becomes painful in any spot, ease the pressure.

• If you work easily and smoothly, your partner will have a warm, tingling feeling in his feet. Did this happen?

• Could you sense the tension in the foot?

Exercise 100 / Foot massage: lying on one's belly

I believe foot massage is the most pleasurable part of any massage. However, some people are overly sensitive in the soles of their feet and cannot tolerate much pressure there. This is due to the spasticity of the muscles there. Massage of these muscles helps relax them, and in time they lose their sensitivity. When this happens, the pleasure of a foot massage increases greatly. If your partner's feet are sensitive to pressure, do the massage gently.

With your partner lying on his belly, do the following:

Place your left hand on the dorsum or upper surface of the foot, and the fist of your right hand against the sole. Rub upward and downward gently. Do the same with your knuckles if your partner wishes it.

Hold the foot in your left hand, and massage each toe with the fingers of your right hand.

Bring the foot upward by having your partner bend the knee. Hold the ankle, and press the palm of your hand against the ball of the foot.

Place both hands on the ball of the foot, and as you press downward, spread the toes gently apart.

Repeat the same maneuvers with the other foot.

• When you finish, place both hands flat against the soles and hold the contact for about a minute.

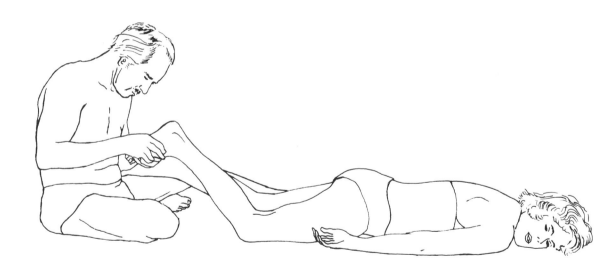

Fig. 65. Foot massage

148

When you withdraw your hands, your partner should still retain an impression of contact for some time. Did this happen?

Exercise 101 / Foot massage: lying on one's back

Have your partner lie on his back on a bed or on a mat on the floor. Sit down by his feet.

Take your partner's left foot in your two hands, and gently stroke the sole of the foot with your thumbs.

Place your left hand under the ankle of his foot and the palm of your right hand against the ball of the foot. Press firmly with the right hand. This flexes and loosens the muscle and provides a good feeling of contact between the foot and the hand.

Place the heels of both hands against the sole, and hold the toes with your fingers. Press with the heels of the hands, and spread the toes with your fingers. This maneuver aims at broadening the foot. Hold the upper surface of the foot with your left hand, and place a fist against the sole.

Rub your fist along the sole using the flat surface of the fist. Then, if your partner can take it, use the knuckles and rub them along the sole.

Take the toes in both hands and bend them downward, but without using too much pressure.

Place your index finger between each toe and massage gently.

Repeat these procedures with the other foot.

Exercise 102 / Back walk

This is a form of massage done with the feet on the back of a person who is lying on a bed or on a mat on the floor. In either case, the bed or mat should be against a wall so that the person who is doing the back walk can balance himself by touching the wall with a hand. Or one can use a chair alongside the mat for balance. This massage maneuver is unique to bioenergetics and is one that most people enjoy very much. It is used in therapy sessions to help patients breathe deeply by relaxing the back muscles. However, persons with lower back problems should not partake in this massage. Nor should the person doing the back walk be too heavy for his partner.

The person to be walked on lies on his belly on a 5'' foam rubber mat or on a bed, legs loosely extended. The person doing the massage should be barefoot. See fig. 66 on next page.

Place one foot crosswise into the

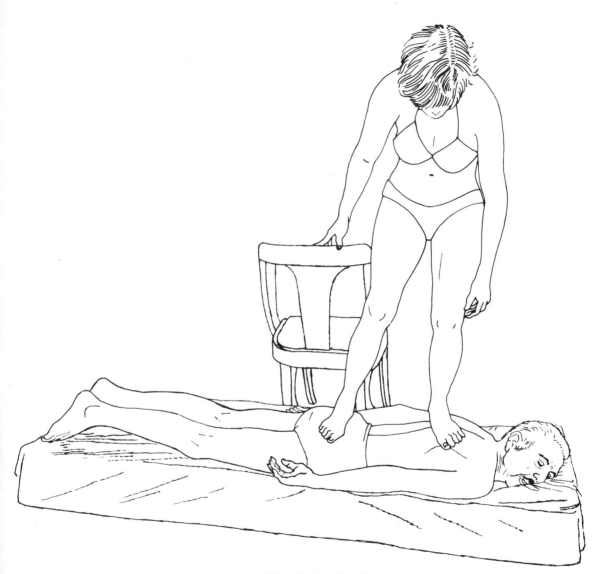

Fig. 66. Back walk

small of the back of your partner and place your other foot on his buttocks, also crosswise. Tell your partner to breathe audibly. As he breathes out, shift your weight to the foot in the small of the back. As he breathes in, shift your weight to the foot on the buttocks. Keep doing this in time with your partner's breathing for about one minute.

While balancing yourself, place one foot on the level of the shoulder blades and the other in the small of the back. Again, shift your weight with the breathing so that the pressure is on the small of the back during the phase of expiration. Continue to do this for about one minute.

If your partner is able to support your weight, place both feet crosswise on the back at the level of the shoulder blades. Tell him to relax and not fight the weight, and he will find that he can breathe fairly well under this pressure.

Walk downward on the back, tak-ing very small steps one foot after another, while your partner relaxes and gives in to your weight. The walk ends when both feet are on the buttocks.

Stand on the buttocks with both feet facing forward and alongside the sacrum. Balancing yourself, bounce up and down rhythmically. This helps shake the pelvis loose and allows the breathing to go deeper.

• Were you aware of your partner's breathing throughout the back walk? There is no undue strain if the person can breathe with your movements.

• Does your partner feel more re-laxed after this procedure?

• Could you sense the tension in your partner's back through your feet?

With some practice, you can be-come very good at this type of back massage and you will find that it will often be requested.

After you have become familiar with the massage techniques described above, you can extend the massage work to include other areas of the body. For example, massaging the calves and thighs is always welcome and appreciated. The basic principle to follow is a sensitivity to your partner. Feel his body and sense which movements make him feel better. Massage is most valuable when it is most pleasurable.

PART III

Setting Up a Regular Schedule

13

Exercises at Home

In chapter 7, we offered some advice about doing these excercises alone and at home. Here we would like to offer some suggestions about the ones that are easiest to do and most productive of good feelings.

Let's say that you only have time to do one exercise a day. Then we recommend that you do exercise 1. This is the exercise in which you bend forward and touch the floor with your fingertips. The knees are slightly bent, and the weight is on the balls of the feet. This exercise is designed to induce vibrations in your legs. If your legs vibrate, you will breathe more easily and more deeply. But whether any vibrations occur or not, this exercise will help you get into your legs and feet and feel more grounded. It is an excellent exercise with which to start the day, but it can be done at any time or place if you feel the need to let down from a state of tension. At most, this exercise takes only about one to two minutes.

If you wish to do two exercises, also do the arch listed as exercise 4 of the series. Start with the arch and hold it for about a minute, breathing into your belly. Then follow it with the forward bend (exercise 1) to get your legs vibrating.

It is much easier to get the legs vibrating if you start with the arch or bow position.

Young people generally have little difficulty getting their legs to vibrate. Older people find it harder because their legs have stiffened with age. One is also stiffer in the morning when one first gets up than later in the day after one has moved about. If this is your condition, we suggest that you do exercise 19 of the standard group after the two preceding exercises. Here, you shift your weight to one leg, bending that knee fully. Allow the weight to rest on that foot until the position becomes uncomfortable, then shift to the other foot. Do this two times on each leg, and you will find that your legs feel more alive and that your center of gravity has dropped. You will feel closer to the ground.

Don't do these simple exercises mechanically. Read the instructions, and pay attention to the questions and comments after each exercise. You are doing these exercises to get in touch with your body, and that requires you to direct your consciousness to what is happening in your body. You also want to get your breathing going, so it is important to keep that function in mind.

If you do more than these three simple exercises regularly, especially in the morning, you will start the day with a better sense of yourself and with more energy. From here on, the choice is open, and it should vary with your feeling, your need, and your time. You may wish to loosen the upper part of your body as part of a regular morning routine. Pick one of the exercises that are given for this part of your body, and add it to your repertoire. You need not do the same exercise every time. After you have familiarized yourself with these exercises, do the ones that help with your particular tensions.

Most people have a lot of tension in their backs. You may be aware of stiffness or rigidity in your back, or you may behave as if you were burdened. The stool is a great help for this kind of problem. We recommend the use of the stool for all people in bioenergetic therapy. In the Appendix there is a list of places where a stool can be obtained. Or you can make one yourself. Try the first exercise given in the section on working with the stool. If you are new to these exercises, place the stool against the bed for security. Feel your way until you gain the sureness that comes with practice. Remember, after using the stool, you should bend forward into the vibrating position.

On the other hand, you may be young, athletic, and in good physical shape. Your interest in these exercises is to see how much more alive your body can feel. You can try all the stool exercises, provided you do not attempt to prove that you are above the need for them and that you can take them in your stride. No one in our culture is free from tension, and since these exercises are designed to get at tension, they will have an effect on you. If you overdo them at first, you may end up suffering distress and pain. Be careful and proceed slowly. If you keep in touch with what is happening in your body, there is no danger in any of the exercises. And, with practice, older people can do them just as well as younger ones.

Most people have a need for greater self-expression. Two of the exercises in chapter 9 on expressive exercises can be done at home with considerable benefit. One is the kicking exercise, and the other is hitting the bed. True, these are exercises that will help develop your aggressiveness and self-assertion, but then these qualities need developing in many people. We have recommended these exercises to our patients for work at home and we do them ourselves. They should be included in your home exercises if you have the time. Read the instructions and comments on these exercises carefully before doing them.

The sexual exercises will *not* solve your sexual problems. They will, however, do much to increase your sexual feelings and your pleasure. You can add any of these exercises to your repertoire, but do them after you have done the preliminary exercises we have recommended here.

It is up to you. The time and energy you invest in your body is the best investment you can make. It pays off in health and pleasure, which are more valuable than money or power. You are really investing in yourself, for you are your body or your body is you. We use the word *exercise* to describe how one works with his own body to help it become or stay more alive. But exercise is a misnomer for these bioenergetic procedures. What you are really doing is caring for your body, and caring involves being interested, feeling, and sensing with affection. Don't exercise your body as if it were a machine or a horse. *Be* your body in its movements, actions, and expressions. That is what this book is about.

14

An Exercise Class

Doing exercises in a group is always more pleasurable and therefore easier than doing them alone. Since these exercises are so extremely helpful to people, we have organized classes that we encourage our patients to attend. Their response has been so positive that we would like to promote these classes wherever bioenergetic therapy is being done.

A class can consist of from four to twenty persons and should have a leader. It is important that the leader be genuinely interested in people and that he or she enjoy doing the exercises. A leader's attitude is picked up by the participants. The leader should also be someone who has a fair amount of knowledge of bioenergetics and has had experience with bioenergetics. His or her role is twofold: to direct the participants properly and to interpret for each person the meaning of his bodily experiences. Another function of the leader is to gauge somewhat the ability of each participant to tolerate some of the stresses involved in the exercises. If a stress seems too great, the participant should be advised to ease off. Nothing is gained by pushing or forcing. Our aim is feeling, not performance.

Because the exercises aim at feeling, it happens frequently that emotions will erupt spontaneously in some members of the class. One of the participants may suddenly break into tears and sobs. Occasionally a participant may become overwhelmed by the new body sensations. In this event, the leader should be sympathetic to the person and keep in touch with what is happening. This may only require the statement, "It's all right. Give in to your feeling and let it out." If the person is upset and disturbed, the leader can go over to her and reassure the person by talking to her. However, unless it is a therapy group, it is inadvisable to try to work out the reasons for the emotional release. That practice would detract from the exercises and leave the other members feeling left out.

It is understood that the leader will personally participate in the exercises as well as direct them. By performing the exercises himself, the leader sets an example for others to follow. At the same time, he should observe the members of the class in order to help them obtain the maximum benefit from the positions used. People cannot see themselves, and quite often they assume they have taken correct positions when in fact they haven't. Only through the proper alignments of the body can one sense the flow of excitation from head to feet.

The leader should make known the purpose of each exercise and the overall goal of the body work. Bioenergetic exercises differ from other types in that they are designed to help the participants let down and give in to the body instead of building muscular strength. The exercises, particularly the stress positions, make you let go of holding or rigidities, with the end result that you feel a little tired physically and outwardly let down but inwardly vibrant, excited, and exhilarated. At the conclusion of a typical exercise class, the participant's breathing should be easier and steadier, his color should be improved, and his eyes should be brighter. The same holds true for the leader. Each should feel more in touch with himself and more grounded.

Before starting a class, each participant should have his medical history checked. Persons not seen by a doctor regularly should have a medical examination. We do not believe that the exercises in this manual, if properly handled, pose any danger, but it is foolhardy to neglect any caution. By the same token, the leader should be alert to any emotional liability in a participant. Some individuals may require special classes where the exercises are geared to their ego strength.

For doing the exercises, try to wear the proper attire. The body has to be somewhat exposed. Women should wear bathing suits or leotards with arms and legs bare; men, shorts or trunks.

One of the values of such a group is that the members can observe each other's bodies and movements. It is easier to see the tensions in another's body than to feel them in yourself. One can thereby gain an understanding of the common tensions that afflict all of us. You also gain support from the others for the commitment to the body work and encouragement when you see someone else improve.

Classes can be private or institutional. It is best if there is some homogeneity in the members of the class. Homogeneity promotes the identification with one another and also allows for more selective use of the exercises. Thus, if you are working with a group of hospital patients, you will avoid the more strenuous and emotional type of exercises in favor of those that focus upon helping the individual become aware of and get in touch with his body. Exercises performed with children or young adolescents in school will be somewhat different from those done with a private group of adults. Such exercise classes have been held (with considerable success), although the authors themselves have not had any experience with these groups. It is the mark of a good leader in bioenergetic exercises that he or she can adapt the exercises to the abilities and needs of the group.

Most exercise classes are coed, although there is considerable value to classes of one sex only. In such uniform groups, especially when they are small and meet regularly, you gain a necessary acceptance from members of your own sex. Very often such classes take on the quality of a group therapy situation. And where these exercises are done as part of a group therapy program, they greatly facilitate the opening up of feeling and communication. Bioenergetic body work brings all people down to the basics of life—breathing, moving, feeling, and expressing.

A class will generally follow the order of exercises given in the standard exercise series. However, it is important to vary the exercises from time to time to avoid the monotony of routine. You can also improvise upon the standard exercises along the basic lines of bioenergetic theory. We have not included all the exercises that we use, and we are constantly devising new ones to meet the needs of the participants or our own needs. Many exercises are created out of self-experience.

Depending on the ego strength of the members of the group, you should include some of the self-expressive exercises. In the hands of a qualified and competent leader, these exercises are powerful tension releasers.

The massage exercises, we believe, should be an integral part of every exercise program if time permits. People enjoy helping one another. The physical contact adds to the closeness and unity of the group. Some people may have objections to touching and being touched, and these objections should be respected by the leader and the other members of the group. These objections are based on fear and will usually fade away in the course of time after they watch others touch and be touched.

An exercise program for an advanced class should include some of the sexual exercises. Unless pelvic tensions are released and sexual feelings opened up, the body cannot regain its natural grace and aliveness. We have repeatedly pointed out that all vibrations will eventually involve the pelvis if they develop fully and lead to spontaneous pelvic movements. To be vibrantly alive is to be sexually alive.

Now, in conclusion, we would like to make a few remarks about the role of these exercises in your life. One doesn't exercise for fun. Exercising can be pleasurable because it makes one feel good, but it is a preventive measure like brushing one's teeth. The consequences of failing to pay attention to the body are serious.

However, these bioenergetic exercises, valuable as they are, are not a substitute for healthy living. To be vibrantly alive, one needs to feel good about one's life, to find some satisfaction in work, and to have pleasure in personal contacts.

You have a responsibility to take care of yourself. You cannot overindulge in food, alcohol, or tobacco and expect the exercises to keep you in vibrant health. A person interested in his body tries to eat sensibly and nutritiously. He tries to get sufficient sleep, to avoid unnecessary stress, and to take time out to breathe and feel himself.

Hopefully, these exercises will enable you to be more "there" as a person in all situations, that is, to be more full-bodied in your responses. We know that they will help you cope better with the stresses of modern living. But doing the bioenergetic exercises at home or in a class is not a substitute for other important physical activities. Being in touch with your body should alert you to the need to engage in other pleasurable body activities. Among these, the simplest is walking. Walk for pleasure, not just to get somewhere. The most pleasurable and exhilarating, apart from sex, is dancing. It is unfortunate we do so little of it. If you are not young and into rock music, we would strongly recommend folk dancing or square dancing. The most healthful body activity is swimming. In swimming, the body is free from the stress of gravity, and breathing and movement have to be coordinated.

The goal of bioenergetic work is to help a person "let go" to pleasure. Pleasure is a bodily response. The capacity for pleasure is a function of body aliveness, that is, how vibrantly alive the body is.

roller —   *Cement Mold Tube*

Appendix

*2 step - top
foam 8 - 10 inches
c̄ tape*

Bioenergetic stools are available at:

New York

Third Avenue Carpenter Shop
346 Third Avenue
New York, New York 10016

Rialto Furniture Company
214 Sullivan Street
New York, New York 10012

Connecticut

John Bellis, M.D.
2804 Whitney Avenue
Hamden, Connecticut 06518

Virginia

Bill Wasserman
Box 49A
Rte. #3
Honaker, Virginia 24260

California   *27 in high $43*

Uniquity — *6*
P.O. Box ~~990~~ *209-745-2111 $40.50*
~~2035 Glyndon Avenue~~ *Galt   CA*
~~Venice, California 90291~~   *95632*

Further information about bioenergetics can be obtained from the bioenergetic centers listed below.

The Institute for Bioenergetic Analysis
144 East 36 Street #1A
New York, New York 10016

Alexander Lowen, M.D., Executive
   Director

714 642-2233
Southern California Bioenergetic Society
833 Dover Drive, Suite #13
Newport Beach, California 92660

Robert Hilton, Ph.D., and Renato
   Monaco, M.D., Trainers

*Kyle Lohmann
Kaj 474-4851*

*Felice Karman
828-3072*

*Margo Roberson
828-8907*

165

Northern California Bioenergetic Society
1307 University Avenue
Berkeley, California 94702

Michael Conant, Ph.D., Trainer

Mid-Western Bioenergetics Center
2791 Walton Boulevard
Rochester, Michigan 48063

Jack McIntyre, M.D.

Tulsa Bioenergetic Center
Tulsa Psychiatric Foundation
1620 E. 12 Street
Tulsa, Oklahoma 74120

Frank Hladky, M.D., Director

Toronto Bioenergetics Center
355 Eglinton Avenue East
Toronto, Ontario, Canada

Kenneth Allen, Ph.D.